Correlation Analysis of Organic Reactivity

With particular reference to multiple regression

CHEMOMETRICS RESEARCH STUDIES SERIES

Series Editor: **Dr. D. Bawden**
Pfizer Central Research, Sandwich, Kent, England

1. Distance Geometry and Conformational Calculations
   G. M. Crippen

2. Clustering of Large Data Sets
   Jure Zupan

3. Multivariate Data Analysis in Industrial Practice
   Paul J. Lewi

4. Correlation Analysis of Organic Reactivity: With particular reference to multiple regression
   John Shorter

Correlation Analysis of Organic Reactivity

With particular reference to multiple regression

John Shorter, M.A., B.Sc., D.Phil., C.Chem., F.R.C.S.
Reader in Physical Organic Chemistry,
University of Hull, England

RESEARCH STUDIES PRESS
A DIVISION OF JOHN WILEY & SONS LTD
Chichester · New York · Brisbane · Toronto · Singapore

RESEARCH STUDIES PRESS

Editorial Office:
58B Station Road, Letchworth, Herts. SG6 3BE, England

Library of Congress Cataloging in Publication Data:

Shorter, John, 1926–
Correlation analysis of organic reactivity
(Chemometrics research studies series; 4)
Includes index.
1. Chemistry, Physical organic. 2. Linear free
energy relationship. 3. Regression analysis.
I. Title. II. Series.
QD476.S44 1982 547.1'3 82-7655
ISBN 0 471 10479 5 AACR2

British Library Cataloguing in Publication Data:

Shorter, John
The correlation analysis of organic reactivity
—(Chemometrics research studies series; 4)
1. Chemical reactions
I. Title II. Series
547.1'3'94 QD505.5

ISBN 0 471 10479 5

Printed in Great Britain

Preface

At the time when I accepted David Bawden's invitation to write a book for the RSP *Chemometrics* Series I had in mind a monograph which would deal almost exclusively with the application of multiple regression in physical organic chemistry. Before I could begin work seriously on this, my book in the Oxford Chemistry Series went out of print. This was *Correlation Analysis in Organic Chemistry: An Introduction to Linear Free-Energy Relationships*, Clarendon Press, Oxford, 1973. It then seemed appropriate that the new book should contain rather more about the elementary aspects of correlation analysis in organic chemistry (CAOC) than I had previously envisaged, and that in consequence the book should be considerably longer than was originally intended. These changes were kindly accepted by RSP. I had already decided to concentrate on chemical reactivity, with only incidental mention of the spectroscopic and biological aspects which were included in the earlier book. These considerations, coupled with the need to find a title for the book which would distinguish it clearly from the earlier book and from other works with which I am associated (see below), have led ultimately to *Correlation Analysis of Organic Reactivity, with particular reference to multiple regression.*

The general plan of the book is explained in Chapter 1. It aims to be a fairly comprehensive guide to the correlation analysis of organic reactivity, with emphasis on the use of multiparametric equations (multiple regression) to obtain more precise correlations and a deeper understanding. It is written at a somewhat higher level than my previous book. I hope, however, that the relative beginner in CAOC will not be entirely lost: the more elementary aspects are outlined and there are plenty of references to guide further reading. I have assumed that readers have a basic knowledge of organic chemistry: nomenclature (both systematic and trivial), characteristic reactions of the common functional groups, reaction mechanisms, and electronic theory. A good grounding in physical chemistry is also assumed: valency theory, kinetics, equilibria, and spectroscopy. I have not assumed any previous knowledge of regression procedures: the treatment begins with a familiar tool of practical work in elementary physics or physical chemistry, *viz.* the equation of the straight line.

It is necessary to say a few words about my presentation of statistical methods. In Chapter 2 I set out the basic ideas and procedures which are necessary for reasonably sound correlation analysis involving simple or multiple linear regression. In Chapter 7 I deal with various statistical matters which seem particularly to concern the practitioners of CAOC. Throughout these two Chapters I have sought to keep the treatment as uncomplicated and informal as possible. I am well aware that in adopting this approach I have rendered myself liable to adverse criticism from those who prefer a more formal and sophisticated approach. However, I am unrepentant: I consider it more fruitful to try and lead the reader gently to an appreciation of the role of correlation analysis in summarizing and rationalizing the data of physical organic chemistry, than to frighten him off with too much emphasis on formal statistical analysis. There are plenty of references to other works in which those who want a more formal treatment can find it.

The correlation analysis of organic reactivity is now such a vast field that it is not possible to deal properly with everything in a book of reasonable size. Inevitably my selection of topics and material reflects my personal preferences and interests, and no doubt there will be those who disagree with my choice. I am well aware that in certain areas I have tended to emphasize examples from the work of our physical organic group at Hull, but perhaps it is best to write about the things of which one has most experience. In the bibliographies I have not tried to mention every worthwhile contribution but altogether I refer to a large number of books, review articles, and original papers, so that the reader may easily pursue in greater detail any topics of particular interest to him. Some critics will consider that I have said too little about the more detailed thermodynamic aspects of CAOC. I had originally hoped to include proper accounts of such topics as the entropy—enthalpy relationship and the reactivity—selectivity principle, but in the end these were excluded by pressures of time and space.

I am very grateful to Oxford University Press for permission to adapt material from my earlier book, and to Plenum Press (New York and London) for permission to draw on my own Chapter in each of the two research monographs which Norman Chapman and I have edited: namely Chapter 2 in *Advances in Linear Free Energy Relationships* (1972), for Chapter 4 in the present book; and Chapter 4 in *Correlation Analysis in Chemistry: Recent Advances* (1978), for Chapter 3 in the present book.

The camera-ready copy of this book has been typed by Janet Bailey of the Chemistry Department. She has been indefatigable in producing work of the highest quality and I am greatly indebted to her. I also thank Maxine Priestman of this Department for her careful drawing of the Figures and structures.

Various academic colleagues have assisted in different ways. Norman Chapman and I have collaborated in research and writing for a quarter of a century and I am glad to acknowledge once again his constant encouragement, helpful advice, and sound criticism. (It is particularly appropriate to refer to our long collaboration at a time when both of us will soon retire prematurely from full-time employment — as a contribution to university economies. We hope, however, that our contributions to physical organic chemistry in various ways will not cease.) I am also indebted in several matters to my colleagues Geoffrey Collier and David Ewing. Ian Pickering, research technician in the Chemistry Department, has done computational work for the book.

Two former Hull colleagues have helped. John Nicholson, now

Professor of Econometrics in the Division of Economic Studies at Sheffield, has acted for many years as my consultant in statistics and he has read and criticized drafts of much of the statistical material in this book. I hasten to affirm that the final responsibility for the presentation of this is mine, for I know he will not approve of everything I have done in the interests of simplicity and informality. Alan Finch, now of Derwent Publications, has for many years alerted me to important papers in CAOC.

I owe a great debt to many friends in the international CAOC community and it seems somewhat invidious to single out any of them for special mention. However, as the bibliographies show, the contributors to the two research monographs mentioned above have provided me with a stimulating background. Marvin and Barbara Charton have often welcomed me to their home and I have been privileged to participate in many 'Brooklyn Heights Meetings' in CAOC. I am also grateful in various ways to Sergio Clementi (Perugia), Otto Exner (Prague), Marek Krygowski (Warsaw), Viktor Palm (Tartu), and Christian Reichardt (Marburg).

Much of the book was drafted while I was on study leave in Oxford, and I thank the Physical Chemistry Laboratory and St. Catherine's College for their kind provision of facilities. I also thank Bill Askew and R.S.P. for their patience and understanding of the circumstances which now tend to hinder the completion of scholarly work by university teachers in this country.

Finally, I make the conventional, but nevertheless sincere, acknowledgment to my wife and family for their support and forbearance. In this I include Hector our dog, who has been deprived of walks, and Zaza our cat, who has often tried to help me by sitting on the manuscript!!

Chemistry Department,
The University,
Hull.

John Shorter

March, 1982

Editorial Foreword

The quantitative treatment of sets of data relating to rates and equilibria of organic reactions, *correlation analysis*, has been one of the most well-established themes of physical organic chemistry. This approach, pioneered by L. P. Hammett in the 1930's, has given much insight into the relationships of chemical structure and reactivity, the selectivity of reactions, and effects of solvent and reagent. It has had applications in the interpretation of spectroscopic data and, particularly significantly, led to the work of Corwin Hansch in the relation of chemical structure to biological activity and to the applications of quantitative structure—activity studies in drug design. Correlation analysis may thus claim to be a major progenitor of the modern field of chemometrics.

In this monograph Dr. Shorter gives a comprehensive introduction to the 'core' of correlation analysis; applications to sets of reaction rate and equilibrium data. Since the book covers the subject from basics, it will be accessible to those with no prior involvement in this area.

The depth of treatment, however, will ensure its value to research workers in this area. In particular it will serve as a source of illustration of the strengths and limitations of the 'classical' multiple regression technique. It is a timely reminder of the continuing usefulness of these methods, given the increasing application of pattern recognition techniques in physical organic chemistry.

David Bawden

Contents

		Page
Chapter 1.	Introduction to correlation analysis in organic chemistry	1
Chapter 2.	Introduction to statistical methods necessary for correlation analysis	9
Chapter 3.	The Hammett equation and its extensions	27
Chapter 4.	The separation of polar, steric, and resonance effects	73
Chapter 5.	Solvent effects	127
Chapter 6.	The role of the reagent	173
Chapter 7.	Some further statistical matters	209
Appendix.	Some terminology of physical organic chemistry	227
Index		231

CHAPTER 1
Introduction to Correlation Analysis in Organic Chemistry

During the last half century chemists have paid increasing attention to the quantitative study of organic compounds and their reactions. Many rate constants and equilibrium constants for organic reactions in solution have been measured and the spectra (infrared, visible, ultraviolet, nuclear magnetic resonance, *etc.*) of thousands of organic compounds have been recorded. The continual improvement in the precision and convenience of instruments is helping to produce an 'information explosion' in this area. As this body of quantitative results has grown it has become increasingly important to have suitable procedures for summarizing and analysing data. The summarizing has involved the development and application of *empirical correlations*, by means of which one body of data can be related to another. At the same time the data may also be analysed to reveal the fundamental factors underlying the phenomena in question. This general approach has become known as *correlation analysis*.

Examples of empirical correlations can be found in the literature of organic chemistry from the turn of the century onwards, but the systematic development of such relationships in the area of organic reactivity began during the years between the two World Wars. Empirical correlations of the reactivities of organic compounds are often linear relationships involving logarithms of rate (k) or equilibrium (K) constants. The first such relationship was the Brönsted equation for general acid or base catalysis (Brönsted and

Pedersen,[1] 1924) but the usual starting point in a discussion of correlation analysis in organic chemistry is the work of Hammett[2] at Columbia and Burkhardt[3] at Manchester in the early 1930s. These authors discovered linear relationships involving log k or log K for a number of systems, notably of the type exemplified in Figure 1.1, in which log k values for the alkaline hydrolysis of

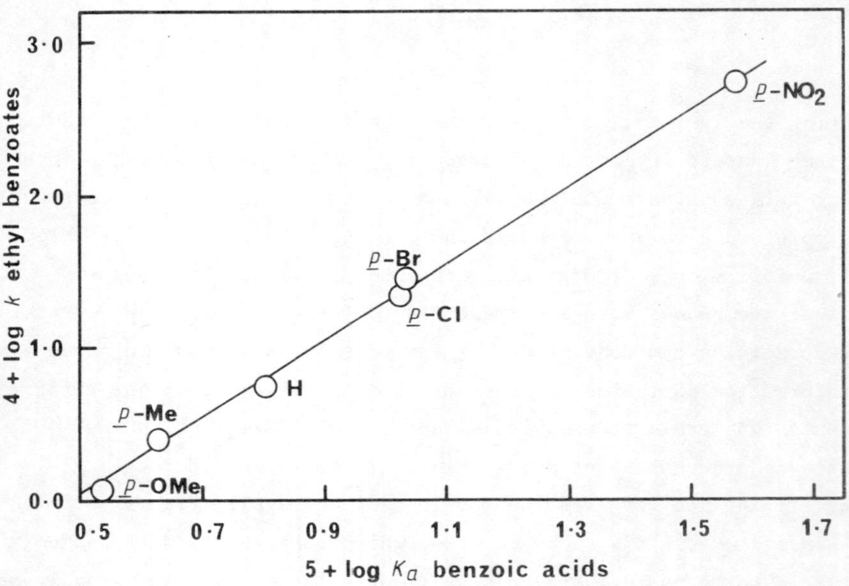

FIG.1.1. Linear free-energy relationship for basic hydrolysis of substituted ethyl benzoates and ionization of substituted benzoic acids. k values (l mol^{-1} sec^{-1}, 85 per cent ethanol w/w, 25°C), C. K. Ingold and W. S. Nathan, *J. Chem. Soc.*, 222 (1936); K_a values, J. F. J. Dippy, *Chem. Rev.*, *25*, 151 (1939).

para-substituted ethyl benzoates are plotted against log K_a values for the ionization of the corresponding substituted benzoic acids. Such a correlation may be described as a rate—equilibrium relationship; we may also have rate— rate or equilibrium—equilibrium relationships.

These relationships are usually referred to as *linear free-energy relationships*. The appropriateness of this term is easily appreciated. For equilibria, equation (1.1) applies:

$$\log K = -\frac{\Delta G^{\ominus}}{2.303\, RT} \quad (1.1)$$

where $\Delta G^{\ominus}$ is the standard free-energy change of reaction. For chemical rate processes:

$$\log k = \log \frac{\mathrm{k}T}{h} - \frac{\Delta G^{\ddagger}}{2.303\, RT} \quad (1.2)$$

where $\Delta G^{\ddagger}$ is the standard free energy of activation. (T is the Kelvin scale temperature; R is the gas constant, k is the Boltzmann constant, and h is Planck's constant.) A relationship between logarithms of K and k (at constant temperature) is thus essentially a relationship between free energies. Each set of rate or equilibrium constants is termed a *reaction series*.

These relationships are so numerous and important that the term linear free-energy relationships (LFER) has often been used to cover the whole field of correlation analysis in organic chemistry. Many of the correlation equations, however, are not LFER in the restricted sense of a relationship involving logarithms of rate or equilibrium constants on both sides of the equation. Further, relationships which are by no means linear are often interesting. It is better to use *correlation analysis* for the whole subject and to use the term LFER more sparingly.

The work of Hammett[2] and of Burkhardt[3] led to the formulation of the Hammett equation which describes the influence of polar *meta*- or *para*-substituents on the side-chain reactions of benzene derivatives.[4] In its original form, and also in some of its refinements, the application of the Hammett equation involves *simple linear regression*. However, in order that the Hammett equation may correlate satisfactorily a wider range of data, it has been transformed into various 'multiparameter extensions', the application of which involves *multiple regression*. All this will be the subject of Chapter 3.

The occurrence of steric as well as polar effects in aliphatic systems and *ortho*-substituted aromatic systems complicates the devising of correlation equations. A typical example is shown in Fig. 1.2. There is clearly no simple relationship between the

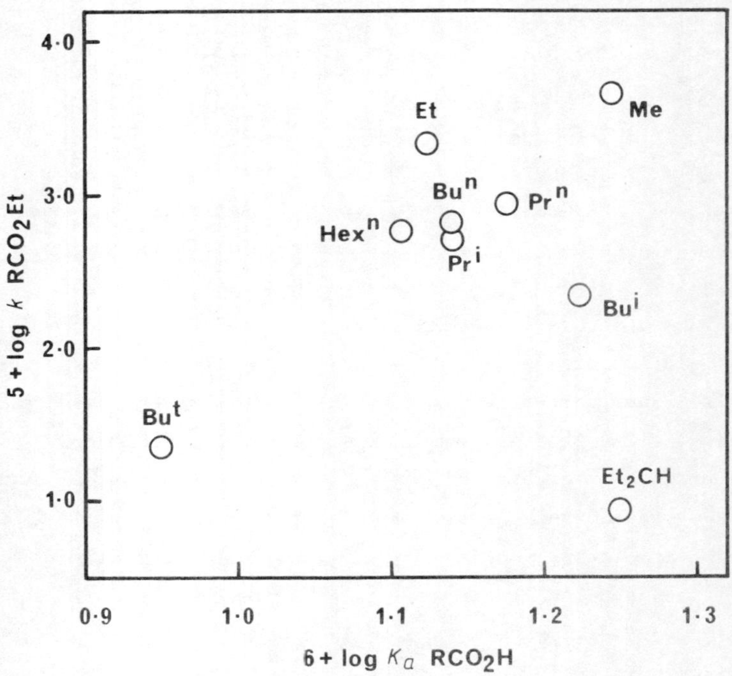

FIG. 1.2. Lack of linear free-energy relationship for basic hydrolysis of ethyl alkanoates and ionization of alkanoic acids. k values (1 mol^{-1} sec^{-1}, 70 per cent acetone v/v, 24.8°C), G. Davies and D. P. Evans, *J. Chem. Soc.*, 339 (1940); K_a values, J. F. J. Dippy, *Chem. Rev.*, 25, 151 (1939).

rates of alkaline hydrolysis of the ethyl esters of the alkanoic acids, and the strengths of the acids themselves. With persistence a few simple LFER *can* be found but no general progress was made until the early 1950s when Taft made an excellent start in satisfactory correlation analysis in this area.[5] This topic is often referred to as "the separation of polar, steric, and resonance effects". As with the Hammett equation, correlation analysis in this area began with simple linear regression, but its further development has involved extensive use of multiple regression. This area is the subject of Chapter 4.

From the work of Hammett and of Taft has come a variety of substituent and reaction parameters of great value for summarizing and understanding the influence of molecular structure on chemical reactivity. The substituent constants have also been applied in fields which are very different from rates and equilibria of organic reactions: to optical spectroscopy, nuclear magnetic resonance spectroscopy, and the mass spectrometry of organic compounds. We shall not be specially concerned with these areas but it will be appropriate to draw from them some examples of the application of multiple regression. With appropriate additional considerations, the fields of enzymology and of structure-activity relationships for drugs have also been greatly illuminated by applying parameters derived from organic chemistry. Although multiple regression is often involved here too, in the limited space available we shall regard these areas as outside the scope of this book.[6]

Organic reactions may be regarded as involving an organic *substrate* reacting with a *reagent* in the presence of a *solvent*. The type of correlation analysis described so far is concerned with reaction series in which the structure of the substrate is varied by introducing substituents, the solvent and the reagent remaining constant. We can also consider the influence of the solvent or the influence of the reagent on the reactivity of a given substrate. Correlation analysis is also valuable here, although much less highly developed and successful. The underlying factors are almost always complex and there is a widespread need to apply multiple regression rather than simple linear regression. This situation has been widely recognized only fairly recently. We consider solvent effects in Chapter 5 (this will include solvent effects on spectroscopic properties) and the effect of the reagent in Chapter 6.

The ultimate value of correlation analysis in organic chemistry lies in the chemical insights which it provides, *i.e.* the way in which it clarifies our understanding of the processes concerned. It cannot do this unless it is on a sound statistical basis. Before we embark on the applications of simple and of multiple regression in physical organic chemistry, we must concern ourselves with the methods of calculating regression coefficients and intercepts, and with estimates of the reliability or success of correlations. Chapter 2 is an introduction to the relevant topics in statistics, to help the reader to understand the meaning of widely-used terms, such as *correlation coefficient,* and to appreciate the steps necessary to make the statistical side of correlation analysis secure, particularly when multiple regression is involved. Certain more specialized matters of statistics will be dealt with towards the end of the book, in Chapter 7. It may then be helpful to refer back to Chapter 2.

An Appendix contains a glossary of the relevant terminology of physical organic chemistry, particularly connected with substituent effects, where there is some variety of practice among different authors.

Throughout this book correlation analysis is presented as regression analysis involving parameter scales based on standard processes. It should be mentioned here that other techniques of multivariate correlation analysis have recently found considerable application to the various aspects of organic reactivity and are exciting increasing interest. These are techniques such as factor analysis, principal component analysis, and pattern recognition, and they do not involve exalting the importance of certain processes as standards. Although these techniques and their applications are strictly outside the scope of this book, limited literature citations are given at appropriate points.[7-10]

REFERENCES

1 J. N. Brönsted and K. Pedersen, *Z. Phys. Chem.*, *108*, 185 (1924).

2 L. P. Hammett and H. L. Pfluger, *J. Amer. Chem. Soc.*, *55*, 4079 (1933).

3 G. N. Burkhardt, W.G.K. Ford, and E. Singleton, *J. Chem. Soc.*, 17 (1936).

4 L. P. Hammett, *J. Amer. Chem. Soc.*, *59*, 96 (1937).

5 Summary by R. W. Taft in *Steric Effects in Organic Chemistry*, M. S. Newman, editor, Wiley, New York, 1956, Chapter 13.

6 For reviews, with extensive bibliographies, covering correlation analysis as applied to the various branches of spectroscopy and biological activity see the appropriate chapters in *Advances in Linear Free Energy Relationships*, N. B. Chapman and J. Shorter, editors, Plenum, London, 1972: and in *Correlation Analysis in Chemistry: Recent Advances*, N. B. Chapman and J. Shorter, editors, Plenum, New York, 1978.

7 S. Wold and M. Sjöström, in *Correlation Analysis in Chemistry: Recent Advances*, N. B. Chapman and J. Shorter, editors, Plenum, New York, 1978, Chapter 1.

8 E. R. Malinowski and D. G. Howery, *Factor Analysis in Chemistry*, Wiley, New York, 1980.

9 P. C. Jurs and T. L. Isenhour, *Chemical Applications of Pattern Recognition*, Wiley, New York, 1975.

10 G. H. E. Nieuwdorp, C. L. de Ligny, and H. C. van Houwelingen, *J. Chem. Soc. Perkin Trans. II*, 537 (1979).

CHAPTER 2
Introduction to Statistical Methods Necessary for Correlation Analysis

		Page
2.1.	Simple linear regression	9
2.2.	Multiple linear regression	19
2.3.	References	24

While the particular concern of this book is with multiple regression it is necessary to start this Chapter with an account of simple linear regression. The Chapter as a whole is intended to provide the necessary background in statistics for a proper appreciation of the subject matter of the book and to enable the reader to make a start on carrying out correlation calculations for himself. A more formal treatment will be found in appropriate books, such as those by Edwards.[1,2] The author has himself profited greatly from the guidance of an econometrician.

2.1. SIMPLE LINEAR REGRESSION

In many experiments in elementary science we are concerned with the effect on a *dependent* variable Y of changes in an *independent* variable X, *i.e.* there is a cause and effect relationship of X to Y. The inspection of a series of corresponding values of these variables may suggest to us that Y is related linearly to X. In this event we plot the graph of Y against X and if the overall pattern of the points seems to agree with the linear relationship, we draw 'by eye' the 'best straight line' through the points. If the scatter of the points about our putative 'best straight line' is commensurate with the experimental errors to which we believe Y and X are subject, we regard the linear relationship as well established, *i.e.* to within the limits of experimental error, the data conform to the equation:

$$Y = aX + b \tag{2.1}$$

where a is the slope of the line and b is the intercept on the Y axis. On the other hand, if the scatter seems out of all proportion to the likely experimental error, then we suppose that there is some complicating factor, possibly another independent variable which is influencing Y.

It is not difficult to find research papers involving correlation analysis in organic chemistry in which the approach is essentially as just described. This is regrettable since the drawing of a best straight line by eye may be highly subjective and the work usually lacks a proper assessment of the precision of the correlation. The slope and the intercept should always be calculated by the *method of least squares*, which also readily provides assessment of the goodness of fit of the data to equation (2.1). However, it is always a good idea to plot and ponder a graph as the first step in a correlation analysis. It may sometimes be dangerous to rush into least squares treatment, particularly in a computerized form. Drawing the graph may reveal deviant points or a curvilinear relationship, which may be less obvious in the output of a computerized least-squares treatment.

Before we deal with the method of least squares, two points of terminology must be mentioned. First, statisticians nowadays often use the term *explanatory variable* rather than independent variable. This term seems to have much to commend it and we shall make some use of it. A second point concerns our use of the term 'correlation' to describe the connection between the dependent variable and the explanatory variable. The statistician uses the term correlation specifically to refer to the strength of the association between two 'random' variables in the population of the two variables, *e.g.* the strength of the linear association between height and weight for a group of eighteen-year old male college students. When the values of an explanatory variable are fixed in controlled experiments and the relationship between the two variables is one of cause and effect, the term *regression* is used. The latter situation is the one with which we deal in correlation analysis in organic chemistry. Hence we use the term regression in the sub-title and frequently in the text of the present book, while we also use the word correlation in its everyday meaning as a general description of what we are doing (as in the title of the present book and in the titles of other works with which the author is associated[3,4]).

As ordinarily carried out, the method of least squares is based on the following convention. It is assumed that the values of the variable X are precisely known and that the experimental error is all in Y; the best straight line is then the line which estimates the values of the dependent variable Y in such a way that the sum of the squares of the deviations between observed and estimated values — the *residual sum of squares*, *RSS* — is a minimum. It should be noted that the *RSS* will contain not only experimental errors but also 'model' errors, *i.e.* errors arising from the inadequacy of the simple linear regression model for the data under analysis. The assumption that the values of X are known exactly is never strictly true for correlation analysis in organic chemistry, although X is often known much more accurately than Y. Strictly speaking, however, least squares treatment in this field should frequently be carried out with more complicated formulae than the simple ones based on the above convention, as presented below. These more general formulae may be found in some books on statistics[5] but they are seldom used in correlation analysis in organic chemistry. On the rare occasions they are applied they usually give results that differ but little from those given by 'improper' application of the simple least squares method. (See the account by S. Wold and M. Sjöström.[6])

If we have n pairs of values of Y and X (data points†), then the *regression coefficient* a and the *intercept term* b are given by

† For this footnote see page 12.

equations (2.2) and (2.3):

$$a = \frac{\Sigma(X - \bar{X})(Y - \bar{Y})}{\Sigma(X - \bar{X})^2} \tag{2.2}$$

$$b = \bar{Y} - a\bar{X} \tag{2.3}$$

where $\bar{X} = \Sigma X/n$, and $\bar{Y} = \Sigma Y/n$, the means of the X and the Y values respectively. If we put $X - \bar{X} = x$ and $Y - \bar{Y} = y$, equation (2.2) may be rewritten very neatly as:

$$a = \frac{\Sigma xy}{\Sigma x^2} \tag{2.4}$$

The simplest measures of the goodness of fit of the regression equation to the data are the *standard deviation of the estimate*, s, and the *correlation coefficient*, r. We give here the expressions for calculating these quantities; their meanings are discussed below.

$$s = \left[\frac{\Sigma y^2 - (\Sigma xy)^2/\Sigma x^2}{n - 2}\right]^{\frac{1}{2}} \tag{2.5}$$

$$r = \frac{\Sigma xy}{\left[\Sigma x^2 \Sigma y^2\right]^{\frac{1}{2}}} \tag{2.6}$$

Very few people any longer work out a, b, s, and r by using logarithms or a simple desk calculator. Numerous programs for large computers are already available for this task and the writing of a simple least-squares program appears as a frequent exercise in computer studies courses. Further, the new programmable desk

Footnote for page 11

† In writing previously about correlation analysis the author has used the term data *sets* instead of data *points*.[7] The term *data set*, however, is now often used in correlation analysis in organic chemistry as synonymous with the term *reaction series* (see Chapter 1). This seems very appropriate, particularly when the correlation analysis of a physical property, rather than chemical reactivity, is involved.

calculators often have a least squares facility. However, if we do attempt to use the above expressions in an old-fashioned way (and this can be of educational value), we must make sure we retain sufficient significant figures right through to the end. Premature rounding-off may lead to serious errors when we are dealing with small differences between large numbers. At the worst we may arrive at a straight line which does not pass between any of the experimental points, or a value of $r > 1$, which is impossible (see below). For calculation by logarithms or simple desk calculator, the expressions for a, b, s, and r may be written more conveniently as follows:

$$a = (\Sigma XY - \Sigma X\Sigma Y/n)/[\Sigma X^2 - (\Sigma X)^2/n] \tag{2.7}$$

$$b = (\Sigma Y - a\Sigma X)/n \tag{2.8}$$

$$s = [\{\Sigma Y^2 - (\Sigma Y)^2/n - (\Sigma XY - \Sigma X\Sigma Y/n)^2/(\Sigma X^2 - (\Sigma X)^2/n)\}/(n - 2)]^{\frac{1}{2}} \tag{2.9}$$

$$r = (\Sigma XY - (\Sigma X\Sigma Y)/n)/\{(\Sigma X^2 - (\Sigma X)^2/n)(\Sigma Y^2 - (\Sigma Y)^2/n)\}^{\frac{1}{2}} \tag{2.10}$$

We deal now with the meaning of s and of r.

s is related to the *RSS*, (*i.e.* the quantity which the method of least squares minimizes, see above) by the equation:

$$s = \left[\frac{RSS}{n - 2}\right]^{\frac{1}{2}} \tag{2.11}$$

$n - 2$ being the *degrees of freedom* of the regression. [s^2 is known as the *residual variance* of Y, *cf.* the *variance* of Y which is $\Sigma(Y - \bar{Y})^2/(n - 1)$. The term $\Sigma(Y - \bar{Y})^2$ may be called the *total sum of squares*, *TSS*, so the variance of Y may be written as $TSS/(n - 1)$.] s is thus a measure of the scatter of the Y values about the regression line. The significance of s is that about two-thirds of the estimated values of Y lie within s units of the corresponding observed values, while about 95 per cent of the estimated values lie within $2s$ thereof. Almost never will an estimate have an error exceeding three standard deviations. If we are interested in the equation as a summary of the data, then the value of s is a means of assessing its precision and is also of use in considering the significance of deviant points.

However, if we are interested in the regression as providing an 'explanation' of the results (*i.e.* as testing the validity of a model), one may reasonably argue that the scatter of the points, which is measured by s, should be looked at in relation to the numerical range of Y covered by the data points. A correlation for which $s = 0.1$ units is highly meritorious if the values of Y cover, say, 100 units but is of little value if Y itself covers only 0.2 units. These considerations are essentially embodied in the correlation coefficient, r, which by combining equations (2.5) and (2.6) may be seen to be related to s by equation (2.12):

$$r = \left[1 - \frac{s^2(n - 2)}{\Sigma y^2}\right]^{\frac{1}{2}} \tag{2.12}$$

Since Σy^2, that is $\Sigma(Y - \bar{Y})^2$, is the total sum of squares, *TSS* (see above), by combining equation (2.12) with (2.11), we have

$$r = \left[1 - \frac{RSS}{TSS}\right]^{\frac{1}{2}} \tag{2.13}$$

$$r = \left[\frac{ESS}{TSS}\right]^{\frac{1}{2}} \tag{2.14}$$

where $ESS = TSS - RSS$, and is the *explained sum of squares*.

Returning to equation (2.12), if $s = 0$, $r = 1.000$ and this defines the perfect regression. As s increases, r decreases and becomes zero when $s^2(n - 2) = \Sigma y^2$. This corresponds to there being no relationship between Y and X, *i.e.* there is no line from which there is less scatter [measured by $s^2(n - 2)$] than there is scatter from the mean value of Y (measured by Σy^2).

Since r is a square-root function any value could be given a positive or a negative sign; r is conventionally taken as lying between 0 and +1 when the regression coefficient a is positive, and 0 and -1 when a is negative. Computers commonly print out correlation coefficients according to this convention. However, in correlation analysis in organic chemistry, the sign of r is generally omitted. Actually the meaningful quantity is r^2, which by equation (2.13) is (1 - *RSS*/*TSS*) or by equation (2.14) is *ESS*/*TSS*, *i.e.* r^2 gives the fraction of the 'variation' of Y (as measured by *TSS*) which is 'explained' by the regression equation.† Thus when $r \approx 0.9$, the equation explains about 80 per cent of the variation of Y.

Following Jaffé (1953)[8] the calculation of correlation coefficients by physical organic chemists is a common practice. There is a much used convention that values of correlation coefficients should be appraised as follows:

0.99 to 1.00	Excellent
0.95 to 0.99	Satisfactory
0.90 to 0.95	Fair
< 0.90	Poor

although sometimes more exacting standards are applied. However any such convention is an over-simplification because correlation coefficients must always be considered in relation to the *number of data points correlated*. This becomes very obvious if we consider the effect of eliminating data points from the regression. The successive elimination of data points will tend always to increase the value of r (unless there is a bias towards leaving in points showing substantial deviations) and in the limit an apparently perfect correlation will be reached when $n = 2$, *i.e.* the minimum number of points necessary to establish the straight line!

Appropriate consideration of r in relation to n may be done by means of the *confidence level* or *significance level* of the correlation. This gives the probability that the correlation could arise by chance from completely unrelated data, *i.e.* the probability

† Since *RSS* and *TSS* are related to the residual variance of Y and the variance of Y respectively (see above), it is sometimes said that r^2 gives the fraction of the *variance* of Y which is explained by the regression equation. This is not strictly true because the expressions for variance and residual variance involve $(n - 1)$ and $(n - 2)$ respectively. The statement holds more exactly as $n \to \infty$, and $(n - 1)/(n - 2) \to$ unity.

of a *null hypothesis* for correlation coefficient being true. This may be expressed as a decimal or a percentage probability. Thus a significance level of 0.001 or 0.1 per cent indicates a probability of one in a thousand, and a significance level of 0.1 or 10 per cent a probability of one in ten, that the correlation could have arisen by chance. A correlation of eight data points giving $r = 0.93$ is significant at about the 0.001 or 0.1 per cent level, while one of three data points giving $r = 0.99$ is significant at about the 0.1 or 10 per cent level. The former correlation, even though it explains only about 87 per cent of the variation of the dependent variable, is securely established, since a probability of one in a thousand can be disregarded. However, the latter correlation, apparently explaining 98 per cent of the variation of the dependent variable, is not securely established, since a probability of one in ten is by no means negligible. In correlation analysis in organic chemistry any apparent correlation at a significance level of 5 per cent or worse should be regarded with suspicion. We should usually expect 1 per cent or better.

It should be noted that significance levels may also be expressed as the differences between 100 and the above percentages, *e.g.* the 0.1 per cent level is then called the 99.9 per cent level. Many physical organic chemists give significance levels in correlation analyses according to the latter convention. It facilitates discussion, in so far as the more significant a correlation is, the larger is the number used to describe its confidence level. Thus we may say that in correlation analysis in organic chemistry we expect correlations to be significant at above the 95 per cent level and usually above the 99 per cent level.

Significance levels may easily be calculated from r and n by using 'Student's t function'. This is:

$$t = r[(n - 2)/(1 - r^2)]^{\frac{1}{2}} \qquad (2.15)$$

[Note that for a given r value, t increases with $(n - 2)^{\frac{1}{2}}$.] From the value of t and of $n - 2$ (the degrees of freedom, sometimes designated as ν), the corresponding significance level can be found from statistical tables[1,9] or from plots of t as a function of $n - 2$ at various significance levels, which are given in certain books, *e.g.* by Moroney.[10] Other authors, *e.g.* Fisher,[11] give tables of values of r for various levels of significance and different numbers of degrees of freedom.

Thus for $n = 8$ and $r = 0.93$, $t = 0.93[6/(1 - 0.93^2)]^{\frac{1}{2}} = 6.20$. At $\nu = 6$, $t = 5.96$ for a significance level of 99.9 per cent, so the actual signifiance level is slightly better than 99.9 per cent.

It is sometimes the practice of physical organic chemists to use the significance level of the correlation coefficient as a criterion of the quality of the regression, to supplement s and r.[12] This is of doubtful validity. Provided that the significance level is, say, better than 99 per cent, it matters little whether it is 99.5 per cent, 99.9 per cent, or 99.999 per cent. The null hypothesis may be rejected emphatically in all these cases, and significance at the 99.999 per cent level does not indicate that a correlation is 'better' than one which is significant at the 99 per cent level.† There is, however, a real need for a criterion which enables the precisions of regressions involving different numbers of data points to be validly compared.

For example, suppose we wish to compare two regressions: (i) $n = 10$, $r = 0.90$, and (ii) $n = 7$, $r = 0.95$. In (i), 81% of the variation of the dependent variable is explained, compared with 90% in (ii), but to what extent is the greater precision of (ii) merely due to the smaller number of data sets? S. Wold has suggested[6] that this question can be answered by comparing the t value in each case with the value corresponding to an acceptable standard level of significance, say 99 per cent, through equation (2.16):

$$C = \frac{t_{1\%}}{t} \qquad (2.16)$$

† Actually the criterion is difficult to apply because the t tables commonly available do not go above the 99.9 per cent level.[13] Many correlations in physical organic chemistry are significant at levels above 99.9 per cent.

If C is <1.0 the significance level of the regression is acceptable, and the closer C is to zero, the better the fit. For (i), $t_{1\%}$ = 3.36, t = 5.85, and C = 0.574; for (ii), $t_{1\%}$ = 4.03, t = 6.80, and C = 0.593. Thus there is little to choose between (i) and (ii); the greater precision of (ii) as measured by r is entirely due to the smaller number of data sets, not to the data being inherently more conformable to the regression model. If, however, the comparison is of (i) with (iii) n = 7, r = 0.98, the greater precision of the latter as measured by r is not solely due to the smaller value of n, for C = 0.367.

It must be mentioned that the regression coefficient a and the intercept term b are random variables with respect to repeated experimentation (*i.e.* of redetermining the value of Y for each fixed value of X) and they have probability distributions and standard errors. The expressions for the standard errors of a and b are:

$$s_a = \frac{s}{\left[\Sigma x^2\right]^{\frac{1}{2}}} \qquad (2.17)$$

$$s_b = s\left[\frac{\Sigma X^2}{n\Sigma x^2}\right]^{\frac{1}{2}} \qquad (2.18)$$

These quantities are often printed out in computerized least-squares calculations. They are useful in a rough and ready way when comparing the a and b values for two related regressions, to see whether there is any significant difference between the respective a and b values. (A strict comparison, however, requires the use of a more complicated statistical procedure.) The ratio a/s_a is equal to the t value for determining the significance level of the regression coefficient, *i.e.* the probability that the value of a could arise by chance from an actual value of zero. The t tests of the null hypotheses for correlation coefficient and regression coefficient are in fact algebraically equivalent. Wold's C criterion discussed above was originally introduced[6] with t in equation (2.16) replaced by a/s_a.

Certain physical organic chemists have devised other measures of precision of simple linear regressions. The best known of these are

Exner's ψ statistic and Taft's f statistic. These will be discussed in Chapter 7.

2.2. MULTIPLE LINEAR REGRESSION

The multiple linear regression of a dependent variable Y with a series of explanatory variables X_1, X_2, X_3, *etc.* in equation (2.19) raises the question of significance in an acute form. Most of our discussion will be about this.

$$Y = a_1X_1 + a_2X_2 + a_3X_3 \ldots + b \qquad (2.19)$$

There are rather complicated equations for applying the method of least squares to the regression of Y on X_1 and X_2, and even more complicated equations for three or more explanatory variables. In correlation analysis in organic chemistry it is rare to undertake regression with more than four explanatory variables. Nowadays multiple regressions are almost invariably performed on a computer, so we will not waste space in stating the necessary equations. In addition to straightforward programs for multiple regression, there are statistical analysis packages which permit the relevance of a large number of possible explanatory variables to be examined ('step-wise regression'). The computer will effect the regressions, starting with a simple linear regression on the most relevant explanatory variable,† and then bringing in the other variables in the order of decreasing relevance, to some pre-determined level of significance. At each stage, values of the intercept term b and of the appropriate regression coefficients a_1, a_2, *etc.*, and values of various statistical quantities, such as standard deviations and correlation coefficients, are calculated, so that the precision of the regression can be assessed through the application of appropriate criteria.

The basic statistical quantities are analogous to those of simple linear regression: they are the standard deviation of the estimate,

† *i.e.* the explanatory variable which gives the highest correlation coefficient in simple linear regression. "Relevance" is tested by correlation coefficient at each stage.

s, and the multiple correlation coefficient, R. Equation (2.20) now defines s,

$$s = \left[\frac{RSS}{n - m} \right]^{\frac{1}{2}} \tag{2.20}$$

where RSS is the *residual sum of squares* (see Section 2.1), n is the number of data points, m - 1 is the number of explanatory variables in the regression, and hence $n - m$ is the number of degrees of freedom, *cf.* the n - 2 term in simple linear regression, equation (2.11). R and s are related by equation (2.21),

$$R = \left[1 - \frac{s^2(n - m)}{\Sigma y^2} \right]^{\frac{1}{2}} \tag{2.21}$$

cf. equation (2.12). Since Σy^2, that is $(Y - \bar{Y})^2$, is the total sum of squares, TSS, by combining equation (2.21) with (2.20), we have

$$R = \left[1 - \frac{RSS}{TSS} \right]^{\frac{1}{2}} \tag{2.22}$$

or

$$R = \left[\frac{ESS}{TSS} \right]^{\frac{1}{2}} \tag{2.23}$$

where $ESS = TSS - RSS$, and is the *explained sum of squares*, *cf.* equations (2.13) and (2.14) in Section 2.1. The general discussion of the meaning of s and of r given in Section 2.1 applies, *mutatis mutandis*, to s and R in multiple regression. Thus s indicates the precision of the regression as a summary of the data, while R^2 gives the fraction of the variation of the dependent variable which is explained by the regression. The significance of R must be assessed in relation to n and m. Increasing the number of explanatory variables in the regression will increase R, but the apparent 'improvement' in any given case may or may not attain a satisfactory level of significance.[14]

The significance level of R may be calculated by using the so-called F distribution, equation (2.24). From the values of F, m - 1 (= ν_1) and $n - m$ (= ν_2), the corresponding significance level can be found from statistical tables.[9]

$$F = \frac{R^2(n - m)}{(1 - R^2)(m - 1)} \qquad (2.24)$$

Thus for $R = 0.93$, $n = 10$ and $m = 3$ (*i.e.* $\nu_1 = 2$ and $\nu_2 = 7$), $F = 22.4$. At these values of ν_1 and ν_2, $F = 9.55$ at the 99 per cent significance level and $F = 21.69$ at the 99.9 per cent level. The actual significance level in this case is therefore very close to 99.9 per cent and is entirely acceptable. On the other hand, for $R = 0.93$, $n = 6$, and $m = 3$ (*i.e.* $\nu_1 = 2$ and $\nu_2 = 3$), $F = 9.61$. At these values of ν_1 and ν_2, $F = 9.55$ at the 95 per cent level and 16.0 at the 97.5 per cent level. Thus the actual significance level is about 95 per cent, which would not be satisfactory.

An acceptable significance level for a final regression involving several explanatory variables is only part of what is necessary. The apparent improvement in correlation coefficient when a new explanatory variable is introduced into the regression must be assessed for significance. This may be done in terms of a *partial correlation coefficient*. Let us take the use of a third explanatory variable as an example. For this purpose the dependent variable is usually designated 1, and the explanatory variables as 2, 3, and 4. The partial correlation coefficient, $r_{14.23}$, for the introduction of variable 4 after 2 and 3 have been used, is given by equation (2.25), where $R_{1.23}$ is the multiple correlation coefficient when 2 and 3 are used, and $R_{1.234}$ is the value when all three explanatory variables are used.

$$r^2_{14.23} = \frac{(R^2_{1.234} - R^2_{1.23})}{(1 - R^2_{1.23})} \qquad (2.25)$$

The denominator gives the fraction of the variation of variable 1 which remains unexplained when variables 2 and 3 are used, and the numerator gives the additional fraction of the variation which is explained by introducing variable 4.

The significance level for the introduction of the additional variable may be evaluated from the t function, which in this case takes the form of equation (2.26),

$$t = r_{14.23}\left[(n - m)/(1 - r^2_{14.23})\right]^{\frac{1}{2}} \qquad (2.26)$$

where $(m - 1)$ is the number of explanatory variables, including the variable whose significance is being examined. Provided that the significance level is adequate, say better than 95 per cent, a partial correlation coefficient of perhaps as low as 0.7 may be deemed acceptable. Thus if $R_{1.23}$ is 0.90, so that about 80 per cent of the variation of the dependent variable is already explained, the explanation of about 50 per cent *of what remains*, making 90 per cent in all, may certainly be of interest.

When several explanatory variables are used it is almost inevitable that there will be some degree of correlation between one 'independent' variable and another. Statisticians call this *multicolinearity*; among physical organic chemists the term *collinearity* is becoming fashionable, as the importance and indeed the dangers of this aspect of multiple regression are increasingly recognized.†

Ideally the various correlation coefficients r_{23}, r_{24}, r_{34}, *etc.* as between the explanatory variables 2, 3, 4, *etc.* should be close to zero. This situation is rarely achieved in practice for the kinds of system with which physical organic chemistry deals, because there is nearly always some physical connection, albeit remote, between the properties that are described by the various explanatory variables. Whenever there is some freedom of choice as to the data for a multiple regression, the aim should be to minimize collinearity. In this matter, however, it does not seem practicable to formulate strict criteria, which should be fulfilled before multiple regression analysis is considered worthwhile. Rather should the limitations of supposed correlations which are subject to the effects of collinearity be clearly recognized and stated.

The effect of collinearity is to increase the standard errors of the regression coefficients, a_1, a_2, a_3, *etc.* (equation 2.19). It is therefore important always to calculate the standard errors s_{a_1},

† The *Concise Oxford Dictionary* gives *collinear*, whereas *Webster's Third New International Dictionary* gives both *colinear* and *collinear*.

s_{a_2}, s_{a_3}, *etc.* and to assess the significance levels of the regression coefficients through the corresponding t values, a_1/s_{a_1}, a_2/s_{a_2}, a_3/s_{a_3}, *etc.* These should be compared with t values at a significance level of at least 95 per cent for $n - m$ degrees of freedom. Further, the *correlation matrix* of r values will instantly reveal the situation regarding collinearity.† Statistical analysis packages for stepwise regression commonly provide all this information. A multiple regression program which yields only the values of the intercept term and regression coefficients, standard deviation of the estimate, and multiple correlation coefficient for an equation involving all the explanatory variables supplied may give very misleading results if collinearities are serious.

This discussion of collinearity prompts finally a mention of certain limitations of correlation analysis. The fact that a dependent variable Y may be well correlated with the explanatory variables X_1 and X_2 does not prove conclusively that the variation of the property measured by Y is to be attributed to the variations in the properties corresponding to X_1 and X_2. There may be other properties of the system, measured, say, by variables X_3 and X_4, which happen to be strongly collinear with X_1 and X_2 respectively, at least within the numerical ranges involved in the data under analysis. Multiple regression of Y on X_3 and X_4 might then give results which were not significantly more or less satisfactory than those of the former regression. Any preference for one regression rather than the other would need to be justified by arguments from outside the particular correlation analysis, *e.g.* that, in relation to the property measured by Y, the physical model corresponding to X_1 and X_2 seems more appropriate than that corresponding to X_3 and X_4.

It is useful to mention what might happen in the above example if stepwise regression of Y on X_1, X_2, X_3, and X_4 were attempted.

† It will not, however, reveal the more subtle form of collinearity: that one explanatory variable is a linear function of two or more of the other explanatory variables.

Let us suppose that X_1 entered the regression first, being marginally preferred to X_3. It could then happen that for setting up the next equation, X_4 would be marginally preferred to X_2. Collinearity might then stop the introduction of X_2 or X_3 into the regression at any worthwhile level of significance. If these variables were forced into the regression the coefficients of X_1 and X_4 would be greatly changed and the standard errors of the regression coefficients for all terms would be considerable, an unsatisfactory situation. Thus the entry of a given explanatory variable into the regression may preclude the participation of any other explanatory variable which is highly collinear with it. So the possible relevance of properties whose quantitative measures are highly collinear cannot be properly examined by multiple regression.[15]

We shall encounter various examples of these problems during the course of this book. In Chapter 7 we shall explore some further statistical aspects of multiple linear regression which appear to concern physical organic chemists, including the use of Exner's ψ statistic and Taft's f statistic mentioned in Section 2.1.

2.3 REFERENCES

1 A. L. Edwards, *An Introduction to Linear Regression and Correlation*, Freeman, San Francisco, 1976.

2 A. L. Edwards, *Multiple Regression and the Analysis of Variance and Covariance*, Freeman, San Francisco, 1979.

3 J. Shorter, *Correlation Analysis in Organic Chemistry*, Oxford Chemistry Series, Clarendon Press, Oxford, 1973.

4 N. B. Chapman and J. Shorter, editors, *Correlation Analysis in Chemistry; Recent Advances*, Plenum, New York, 1978.

5 O. L. Davies and P. L. Goldsmith, editors, *Statistical Methods in Research and Production*, Oliver and Boyd, Edinburgh, 4th edn., 1972.

6 S. Wold and M. Sjöström, Chapter 1, in ref. 4.

7 Appendix in ref. 3.

8 H. H. Jaffé, *Chem. Rev.*, *53*, 191 (1953).

9 *e.g.* D. V. Lindley and J.C.P. Miller, *Cambridge Elementary Statistical Tables*, C.U.P., 1966.

10 M. J. Moroney, *Facts from Figures*, Penguin Books, Harmondsworth, 1956.

11 R. A. Fisher, *Statistical Methods for Research Workers*, Oliver and Boyd, Edinburgh, 14th edn., 1970.

12 See Appendix in ref. 3. (Regrettably!)

13 See, however, Appendix Table IV in ref. 1.

14 The 'improvement' accomplished by additional explanatory variables is an issue of special importance in the correlation analysis of biological activity. See J. G. Topliss and R. P. Edwards, *J. Med. Chem.*, *22*, 1238 (1979).

15 The *Chemometrics* Series editor, David Bawden, has pointed out to the author that this is a situation in which principal component analysis is most advantageous.

CHAPTER 3

The Hammett Equation and its Extensions

Page

3.1. Introduction 27

3.1.1. The sliding-scale of sigma values 32

3.2. The Yukawa-Tsuno equation 35

3.2.1. The significance of $r^{\pm}$ 37

3.2.2. The problem of two-step processes 40

3.2.3. Other complications 41

3.2.4. Electrophilic aromatic substitution 42

3.2.5. A recent exchange of views 44

3.3. Taft's dual substituent-parameter equation 45

3.3.1. The interpretation of ρ_I and ρ_R values 51

3.3.2. The scaling of σ_I and σ_R values 52

3.4. The Swain—Lupton treatment 54

3.4.1. Applications of the Swain—Lupton treatment 58

3.4.2. Critique of the Swain—Lupton treatment 59

3.4.3. Extensions of the Swain—Lupton treatment 60

3.5. Miscellaneous topics and some recent developments 62

3.6. References 68

3.1. INTRODUCTION

As mentioned in Chapter 1 various linear free-energy relationships were discovered in the early 1930s for the side-chain reactions of *meta*- or *para*- substituted benzene derivatives. Hammett's signal contribution in 1937[1,2] lay essentially in recognizing the value of taking one reaction as a standard process, with which all other relevant reactions could be compared. Much information about reactivity could thereby be summarized in terms of a very simple mathematical equation, with its characteristic constants. The concept of the 'standard reaction series' has dominated the development of correlation analysis in organic chemistry. As the subject has expanded to take in a wider variety of data and as the analysis has been made more precise, the number of standard processes has grown.

Equations (3.1) and (3.2) show the basic forms of the Hammett equation, in which k or K is the rate or equilibrium constant for a

$$\log k = \log k^{o} + \rho\sigma \qquad (3.1)$$

$$\log K = \log K^{o} + \rho\sigma \qquad (3.2)$$

side-chain reaction of a *meta*- or *para*- substituted benzene derivative. The symbol k^{o} or K^{o} denotes the statistical quantity (intercept term) approximating to k or K for the 'parent' or 'unsubstituted' compound. The *substituent constant*, σ, measures the polar (electronic) effect of replacing H by a given substituent (in the *meta*- or *para*- position) and is, in principle, independent of the nature of the reaction. The *reaction constant*, ρ, depends on the nature of the reaction (including conditions such as solvent and temperature) and measures the susceptibility of the reaction to polar effects. Hammett chose the ionization of benzoic acids in water at 25°C as the standard process. For this, ρ is defined as 1.000, and the value of σ for a given substituent is then $\log (K_a/K_a^{o})$, where K_a is the ionization constant of the substituted benzoic acid and K_a^{o} that of benzoic acid itself.

To apply the Hammett equation to a given reaction, rate or equilibrium data must be available for the parent compound and several (preferably three or more) *meta*- or *para*- substituted compounds involving substituents of known σ. Log k or log K, as appropriate, is plotted against σ, the parent compound providing a data point at $\sigma = 0$. The best straight line should be obtained, however, by the method of least squares. Log k or log K is taken as the dependent variable and σ as the explanatory variable, and values of ρ (the slope) and log k^{o} or log K^{o} (the intercept at $\sigma = 0$) are calculated, (see Chapter 2).†

The success of Hammett correlations is commonly assessed (following Jaffé[3]) by calculating the correlation coefficient, r, and the standard deviation, s, of the estimated values of log k or log K. (See Chapter 2, including certain *caveats*. Exner's statistic ψ, and the standard deviation of ρ, s_{ρ} are also useful measures of success. See also Chapter 7.)

Values of σ for certain common substituents are given in Table 3.1. Electron-withdrawing substituents have positive values of σ and electron-releasing substituents have negative values. For more extensive compilations see Exner,[4] and Hansch and Leo,[5] and for the interpretation of σ values in simple electronic terms see Shorter,[6] Hine[7] and Johnson.[8] We defer any tabulation and discussion of ρ values until we deal with the multiparameter extensions of the Hammett equation later in this chapter. (See also discussions by Shorter[6] and by Johnson.[8])

It is not easy to assess what proportion of the relevant reactions conform reasonably well to the Hammett equation in the simple form

---

† The form of the equation is subject to minor variations, of which the most serious is the combining of the log terms as log (k/k^{o}) or log (K/K^{o}), with k^{o} or K^{o} now referring to the *observed* value for the parent system. This procedure is, of course, necessary in the definition of σ from benzoic acid ionization but it is not a necessary general procedure to force the regression through the point for the parent system. It is preferable to confirm that the intercept term is indistinguishable from the experimental value for the parent system and thus check that there is no effect connected with the replacement of H by *any* other substituent. If only *relative reactivities* are available, then log (rel. react.) is the dependent variable in the regression on σ, but the regression should still not be forced through the origin. An intercept should be permitted in the least squares treatment but in the ordinary way should emerge as indistinguishable from zero. (See Section 7.1.)

TABLE 3.1. Selected values[a] of σ

Substituent	σ_m	σ_p	Substituent	σ_m	σ_p
Me	-0.07	-0.16	OH	0.13	-0.38
Bu^t	-0.10	-0.20	OMe	0.11	-0.28
Ph	0.06	-0.01	OAc	0.39	0.31
CF_3	0.43	0.54	SH	0.25	0.15
COMe	0.38	0.50	SMe	0.15	0.00
CO_2H	0.37	0.45	$SO_2{\cdot}Me$	0.60	0.72
CN	0.61	0.70	F	0.34	0.06
NH_2	0.00	-0.57	Cl	0.37	0.22
NMe_2	-0.15	-0.63	Br	0.39	0.22
NO_2	0.71	0.78	I	0.35	0.21

a From the compilation by Exner (1978).[4] Mainly based directly on the ionization of substituted benzoic acids. Some of the values differ slightly from those given in the earlier compilation by D. H. McDaniel and H. C. Brown, *J. Org. Chem.*, *23*, 420 (1958). The values should not be used uncritically; see estimates of accuracy and the various admonitions in the compilations. For some substituents there are significant differences between values determined by various authors.

---

described above. Jaffé (1953)[3] examined its application to about 400 reaction series, laying great stress on correlation coefficient as a measure of the success of the Hammett equation. It is now held that insufficient attention was paid to the effect of degrees of freedom in regression (see Chapter 2), but whatever the limitations of the detailed statistics may be, the general conclusions to be drawn from Jaffé's review are quite clear. First, a very considerable body of data conforms quite well to the Hammett equation, with a mean precision of about 15 per cent in k or K, which is a not very exacting standard in relation to the precision with which many rate or equilibrium measurements have been made. Secondly, however, there is a substantial minority of rate or equilibrium data that is outside the scope of the Hammett equation in its original form and mode of application. Hammett himself had found that p-NO_2 showed deviations in the correlation of reactions of anilines and phenols.[1,2] The deviations were systematic in that a σ value of *ca.* 1.27 seemed to apply, compared with 0.78 based on the ionization of *p*-nitrobenzoic acid. The gradual discovery of similar examples pointed the way to a first refinement of the Hammett equation through a 'duality of substituent constants'.[6,8]

When σ values based on the ionization of benzoic acid are used, deviations may occur with +*R* *para*-substituents for reactions involving -*R* electron-rich reaction centres, and with -*R* *para*-substituents for reactions involving +*R* electron-poor reaction centres. (See Appendix for exposition of signs and symbols.) The explanation of these deviations is in terms of 'cross-conjugation', *i.e.* conjugation involving substituent and reaction centre, which is substantially more marked than in the ionization of benzoic acid.[6,8] In each case 'exalted' σ values are required for conformity to the Hammett equation. The special constants for +*R* substituents (*e.g.* NO_2, CN, CO_2H, CO_2Me, and SO_2Me) are denoted by σ^-, and those for -*R* substituents (*e.g.* OMe, Me, OH, NH_2, SMe, and Hal) are denoted by σ^+. New standard series are necessary to establish the scales of σ^- and σ^+. The former may be based either on the ionization of anilinium ions or of phenols in water. (For some substituents the values based on the two systems differ slightly.) In either case the Hammett line is established with *meta*-substituents and -*R* *para*-substituents. Values of σ^+ were based by Brown and Okamoto (1958)[9] on the solvolysis of t-cumyl chlorides in 90% acetone-water at 25°C, the Hammett line being established with *meta*-substituents and +*R* *para*-substituents. (In establishing the Hammett lines for σ^-, strong +*R* substituents are not used in the *meta*-position, and for σ^+ strong -*R* substituents are similarly not used, in case there is a relayed resonance effect. For such substituents σ_m^- or σ_m^+ values may subsequently be obtained. Brown obtained σ_m^+ or σ_p^+ for *all* substituents from the Hammett line, but the validity of the small corrections applied to substituents already used to establish the line is doubtful.) A selection of σ^+ and σ^- values is in Table 3.2. See Exner[4] for larger compilations.

Values of σ^- have been applied extensively to the correlation of data for the reactions of aniline and phenol derivatives, of

TABLE 3.2. Selected values[a] of σ_p, σ_p^+, and σ_p^-

Substituent	σ_p	σ_p^+	Substituent	σ_p	σ_p^-
Me	-0.16	-0.31	CF_3	0.54	(0.74)
Bu^t	-0.20	-0.26	CN	0.61	0.88
Ph	-0.01	-0.18	Ph	-0.01	0.11
NH_2	-0.57	(-1.4)	COMe	0.50	0.84
OH	-0.38	(-0.92)	$CONH_2$	0.31	0.61
OMe	-0.28	-0.78	CO_2H	0.45	(0.78)
F	0.06	-0.07	$SiMe_3$	0.00	0.07
Cl	0.22	0.11	NO_2	0.78	1.28
Br	0.22	0.15	$SO_2{\cdot}Me$	0.72	0.98
I	0.21	0.13	$SO_2{\cdot}NH_2$	0.57	(0.89)

a From the compilation by Exner.[4] σ_p^+ based on t-cumyl chloride solvolysis, with exception of NH_2 and OH. σ_p^- based on ionization of phenols, with exception of CF_3, CO_2H, and $SO_2{\cdot}NH_2$ (ionization of anilinium ions).

carbanions, and of radicals, and have also been applied to nucleophilic aromatic substitution, *i.e.* where there is no side-chain and reaction occurs at a ring carbon. Values of σ^+ have been applied extensively to solvolyses, aldehyde reactions, basicities of carbonyl compounds and amides, and radical reactions, and most notably to electrophilic aromatic substitution.

3.1.1. THE SLIDING-SCALE OF SIGMA VALUES

The use of σ^+ and σ^- greatly extends the range of applicability of the Hammett equation. There are, however, some situations in which the use of σ is unsatisfactory, but the use of σ^+ or σ^- (as appropriate) appears to over-correct for the effect of cross-conjugation. Indeed the 'duality of substituent' constants and the attempt to deal with cross-conjugation by selecting σ^+, σ, or σ^- in any given case is somewhat artificial. The contribution of the resonance effect of a substituent relative to its inductive effect might be expected to vary continuously as the electron-demanding quality of the reaction centre is varied, *i.e.* the extent to which it is electron-rich or electron-poor. Thus a *sliding-scale* of substituent constants might be expected for each substituent having a resonance effect, and not just a pair of discrete values: σ^+ and σ for $-R$, or σ^- and σ for $+R$ substituents.

The problem of putting into practice the idea of the sliding-scale exercised various authors in the late 1950s. Either the Hammett equation had to be used in a different way or it had to be modified.[‡] Several types of treatment were devised, mainly involving multiparameter extensions of the Hammett equation and hence the need to apply multiple regression.

Before we discuss the multiparameter extensions of the Hammett equation, we must mention the work of van Bekkum, Verkade, and Wepster (1959),[11] who strongly criticized the duality of sigma constants as an artificial device to describe "mesomeric *para*-interaction", as they termed cross-conjugation. They proposed a new method of applying the Hammett equation. In this method the ionization of benzoic acids is used only to evaluate 'primary' σ values which can be considered unambiguously 'normal', *i.e.* free from the effects of cross-conjugation. Only such σ values are used in calculating ρ for a given reaction series. For other substituents the Hammett line is then used to obtain σ values relevant to the particular reaction series. The authors applied this method to about eighty reaction series and confirmed the hypothesis of a sliding-scale of sigma values arising from cross-conjugation. From the multiplicity of sigma values recorded for each substituent, the authors selected a normal (unexalted) value, designated σ^n. It was assumed that $-R$ reaction centres give normal sigma values for $-R$

---

‡ It has to be recognized that developments along these lines carried within themselves the seeds of the devaluation of this system of empirical correlation, through the proliferation of sigma scales and the increasing complexity of the analysis. In the limit the 'substituent' constant might become characteristic of the reaction as well. Some authors[10] believe that what has happened in the field in the last 20 years indicates that these worst fears have been realized. However a greater understanding of structure— reactivity relationships has undoubtedly emerged from the welter of discussion.

TABLE 3.3. Selected values[a] of σ, σ (primary), and σ^n

Substituent	σ	σ (primary)	σ^n
m-Me	-0.07	-0.07	-
m-COMe	0.38	0.38	-
m-NH_2	-0.00	-	-0.04
m-NO_2	0.71	0.71	-
m-OH	0.13	-	0.10
m-OMe	0.11	-	0.08
m-F	0.34	0.34	-
m-Cl	0.37	0.37	-
m-Br	0.39	0.39	-
m-I	0.35	0.35	-
p-Me	-0.16	-	-0.13
p-COMe	0.50	0.50[b]	-
p-NH_2	-0.57	-	-0.17
p-NO_2	0.78	0.78[b]	-
p-OH	-0.38	-	-0.18
p-OMe	-0.28	-	-0.11
p-F	0.06	-	0.06[c]
p-Cl	0.22	-	0.24
p-Br	0.22	-	0.27
p-I	0.21	-	0.30

a σ values from Exner;[4] values of σ (primary) and σ^n from van Bekkum *et al*.[11]

b Used for reactions where cross-conjugation does not occur.

c Regarded as suspect by van Bekkum *et al*.;[11] they considered the proper value should be *ca*. 0.17.

substituents, and +*R* centres for +*R* substituents, a small spread of calculated σ^n values being averaged for each substituent (see Table 3.3).

For calculating σ, eight primary σ values of *meta*-substituents (including H, $\sigma = 0.00$) are of general applicability, and *p*-COMe and *p*-NO_2 are considered acceptable for systems in which cross-conjugation can be ruled out (see Table 3.3). All other *para*-substituents are unacceptable, along with *meta*-substituents such as OH and OMe, whose strong -*R* effects may cause anomalies even from the *meta*-position. The authors claimed that these restrictions led to striking improvements in the success of Hammett correlations, but this conclusion was not entirely secure because insufficient attention was paid to the increase in correlation coefficient due to reducing the number of degrees of freedom of the regression (see Chapter 2). There is, however, no doubt that the authors' main chemical ideas were sound.

This work of van Bekkum *et al.*[11] was important but it tends towards the situation in which the 'substituent constant' becomes reaction dependent, and correlation analysis is devalued (see footnote ‡ earlier). Therefore we turn now to multiparameter extensions of the Hammett equation, of which there are three principal forms: (i) the Yukawa—Tsuno equation, (ii) Taft's dual substituent-parameter equation, (iii) the Swain—Lupton treatment. We shall deal with these in turn.

3.2. THE YUKAWA—TSUNO EQUATION

Yukawa and Tsuno (1959)[12] proposed a method for dealing with the influence of -*R* substituents on reactions that are more electron-demanding than the ionization of benzoic acid. They suggested that values of $(\sigma^+ - \sigma)$ would provide a scale of enhanced resonance effects and they modified the Hammett equation to incorporate this feature. There is a considerable variety of practice as to details of form and symbolism for the Yukawa—Tsuno equation. We will not concern ourselves with these details but will simply write the equation in a way which is convenient for the present discussion, in which the enhanced resonance effect $(\sigma^+ - \sigma)$ is written as $\Delta\sigma_R^+$. (A corresponding equation for equilibria may also be written.) The equation implies

$$\log k = \log k^o + \rho(\sigma + r^+\Delta\sigma_R^+) \qquad (3.3)$$

the multiple linear regression of log k on σ and $\Delta\sigma_R^+$. (See footnote in Section 3.1 concerning the practice of combining the log terms.) However, for $+R$ *para* - substituents and for all *meta*-substituents except those of powerful $-R$ effect, simple linear correlation with σ is effectively involved. The quantity r^+ gives the contribution of the enhanced resonance effect of $-R$ substituents. If $r^+ = 0$ the equation reduces to the simple Hammett equation and if $r^+ = 1$ it corresponds to straightforward correlation with σ^+. Values of $r^+ > 1$ are also conceivable if there are reactions that are more electron-demanding than the solvolysis of t-cumyl chlorides (used to define σ^+).

A corresponding equation (3.4) with σ^- constants to deal with the influence of $+R$ substituents on reactions that are less electron-demanding than the ionization of benzoic acid was formulated by Yoshioka *et al.* (1962).[13]

$$\log k = \log k^o + \rho(\sigma + r^-\Delta\sigma_R^-) \qquad (3.4)$$

The name Yukawa— Tsuno equation is commonly used as a collective description of both the equations, and also of the modified equations which use Taft's 'normal' substituent constant σ^o, which is analogous to σ^n (see above), but is derived differently (see Section 3.3).

$$\log k = \log k^o + \rho(\sigma^o + r^+\Delta\bar{\sigma}_R^+) \qquad (3.5)$$

and

$$\log k = \log k^o + \rho(\sigma^o + r^-\Delta\bar{\sigma}_R^-), \qquad (3.6)$$

where $\Delta\bar{\sigma}_R^+ = (\sigma^+ - \sigma^o)$, and $\Delta\bar{\sigma}_R^- = (\sigma^- - \sigma^o)$.[14] These forms of the equation may be held to be theoretically more correct than the original forms (equations 3.3 and 3.4) since the σ^o scale is a more proper origin from which to assess enhanced resonance effects. The original forms, however, are still often used by certain authors.†

† The symbol r rather than r^+, r^- or $r^\pm$ has often been used for the characteristic Yukawa— Tsuno coefficient,[12] but there is possible confusion with r meaning correlation coefficient.[15]

We shall not attempt here to survey the innumerable applications of the various forms of the equation (see Shorter[15]) but rather to deal with various issues regarding the validity and meaning of Yukawa–Tsuno correlations. It would be appropriate, however, to list a few of the areas in which the equation has been extensively applied and to draw on selected examples for illustration. The areas of application include: solvolyses, electrophilic aromatic substitution (including protodemetallation and protodetritiation), acid-catalysed hydrolyses of diazocompounds, nucleophilic reactions, additions to multiple carbon-carbon bonds, and physical data such as chemical shifts and ir frequencies. Innumerable claims have been made that the Yukawa–Tsuno treatment is more successful than simple correlation with σ^+, σ, or σ^-; the $r^\pm$ values obtained have often been interpreted in mechanistic terms.

Yukawa and Tsuno[12] suggested two methods of correlation analysis with their equation. The most general may be indicated by rewriting equation (3.3) in the form:

$$\log k = \log k^o + a\sigma + b\Delta\sigma_R^+ \qquad (3.7)$$

Multiple regression of log k on σ and $\Delta\sigma_R^+$ gives a (= ρ) and b (= $r^+\rho$), and log k^o is the intercept corresponding to log k for the parent system. Alternatively, when adequate data for suitable *meta*-substituents are available, ρ may be determined by simple regression of log k on σ, $\Delta\sigma^+$ being zero. (This is essentially the procedure of van Bekkum *et al.*[11]) Simple regression of (log k - $\rho\sigma$) on $\Delta\sigma_R^+$ for the remaining substituents then gives b and log k^o.

The equation has sometimes been applied by plotting log k *versus* ($\sigma + r^+\Delta\sigma_R^+$) for trial values of r^+ to see which plot gives the 'most straight' line, the plot against σ curving in one direction and that against σ^+ curving in the other.

3.2.1. THE SIGNIFICANCE OF $r^\pm$

Along the lines suggested originally by Yukawa and Tsuno,[12] $r^\pm$ is usually supposed to measure the extent to which cross-conjugation of substituent with reaction centre stabilizes the transition state (or products in the case of equilibria) relative to the initial state, on

TABLE 3.4. Applications of the Yukawa–Tsuno equation[a]

Reaction	ρ	r^+	s^b	n^c
Beckmann rearrangements:				
(1) Acetophenone oxime picryl ethers, 1,4-dichlorobutane, 70°C	-4.24	0.46	0.233	12
(2) Acetophenone oximes, 94.5% sulphuric acid, 50.9°C	-1.98	0.43	0.044	9
Solvolysis of t-cumyl chlorides, 90% aq. acetone, 25°C	-4.52	1.00	0.093	21
Acetolysis of neophyl brosylates, 70°C	-3.38	0.54	0.077	7
Protodemetallation of $XC_6H_4 \cdot MR_3$ (various conditions):				
M = Si	≈ -5.1	≈ 0.65	-	18
Ge	≈ -4.4	≈ 0.6	-	19
Sn	≈ -3.8	≈ 0.4	-	13
Pb	≈ -2.5	≈ 0.4	-	5
Addition of Br_2/0.2M KBr MeOH, 25°C to				
(1) monosubstituted 1,1-diphenylethylenes	-3.57	0.84	0.019	10
(2) substituted styrenes	-4.30	1.0	0.156	6
	ρ	r^-	Correlation coefficient[d]	n^c
Alkaline hydrolysis, 56% w/w acetone–water, for				
(1) aryl acetates, 1°C	1.69	0.2	0.985	16
(2) aryl benzoates, 25°C	1.76	0.3	0.995	12
Reaction of picryl chloride, in 75% v/v aq.EtOH, 30°C, with				
(1) phenoxide ions	-2.27	0.94	0.96	14
(2) anilines	-3.57	0.68	0.997	14

a scale established by taking $r^{\pm}$ as unity or zero for certain standard processes. Many authors have interpreted $r^{\pm}$ values for particular reactions on this basis. Further it may be argued that $r^{\pm}$ is a better indicator than ρ of the structures of initial states, transition states, and products as appropriate, since ρ is a more complex quantity, being influenced by all the interactions between substituents and reaction centre, *e.g.* electrostatic effects. (There is, in fact, no general relationship between ρ and $r^{\pm}$ to be discerned over any wide and varied range of reactions. See Shorter.[15]) We will give just a few examples in which the value of $r^{\pm}$ appears to yield useful information. The data to be discussed are in Table 3.4.

The ρ values for the two Beckmann rearrangements in Table 3.4 differ greatly, but the r^{+} values are similar, which may well indicate a similarity of transition states, which is independent of leaving group and reaction conditions. On the other hand, the different r^{+} values for the solvolysis of t-cumyl chlorides and of the neophyl brosylates may indicate somewhat different modes of resonance stabilization for the incipient carbonium ions in the two transition states [see (I) and (II)].

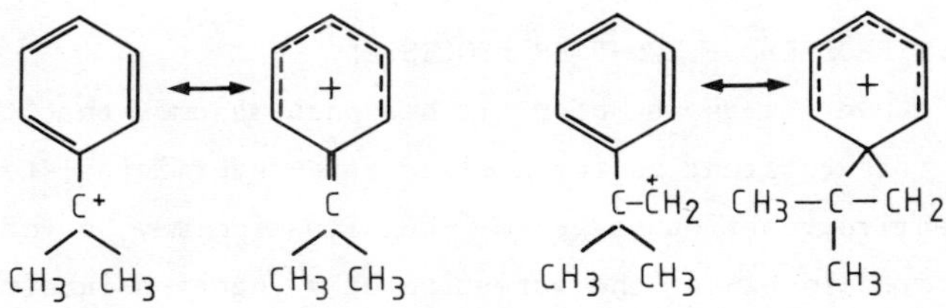

(I) (II)

---

a References to the original literature may be found *via* ref. 15.

b Standard deviation of the estimate.

c Number of data points.

d For simple regression on $[\sigma + r^{-}(\sigma^{-} - \sigma)]$.

The finding of higher values of r^+ for protodesilylation and protodegermanylation than for protodestannylation and protodeplumbylation has been taken to indicate that the electrons of the more polarizable Ar-Sn and Ar-Pb bonds help to delocalize positive charge in the transition state, so that there is less demand on the electrons of the ring.

The different r^+ values for the addition of bromine to monosubstituted 1,1-diphenylethylenes and to substituted styrenes show that the two rings of 1,1-diphenylethylene cannot be simultaneously coplanar with the olefinic bond, so resonance interaction with the carbonium centre in the intermediate is reduced.

The small r^- values for the alkaline hydrolysis of aryl acetates and benzoates indicate only a small extent of bond-breaking (release of phenoxide ion) in the transition state. On the other hand, in the reaction of picryl chloride with phenoxide ions and with anilines the high r^- values show that considerable bond formation has occurred in the relevant transition state.

Such examples could be multiplied greatly but it is necessary now to point out that there are certain difficulties, both theoretical and practical, which beset the application of the Yukawa—Tsuno equation and reduce confidence in the significance of $r^{\pm}$ as often measured.

3.2.2. THE PROBLEM OF TWO-STEP PROCESSES

In the above discussion of $r^{\pm}$ it has been assumed that the observed rate constant is for a fixed rate-determining step throughout the entire reaction series. The situation may be very different for reactions involving the formation of a reaction intermediate. This matter has recently been reviewed by Shorter[15] and we must confine ourselves here to a few main points. Reaction intermediates are often formed in a pre-equilibrium followed by a rate-limiting step:

$$\mathrm{A} + \mathrm{B} \underset{}{\overset{K}{\rightleftharpoons}} \mathrm{I} \xrightarrow{k_2} \text{products} \qquad (3.8)$$

The observed rate constant is then given by $k = Kk_2$. If the simple Hammett equation is obeyed, $\rho = \rho_1 + \rho_2$, where ρ_1 is for the pre-equilibrium and ρ_2 for the rate-limiting step. In a reaction series to which the Yukawa—Tsuno equation is apparently applicable, $r^{\pm}$ as observed will be related to $r_1^{\pm}$ and $r_2^{\pm}$ for the pre-equilibrium and rate-limiting steps, respectively, by the equation:

$$r^{\pm} = \frac{\rho_1 r_1^{\pm} + \rho_2 r_2^{\pm}}{\rho_1 + \rho_2} \qquad (3.9)$$

In such a case $r^{\pm}$ has no clear physical significance. Therefore one must view with scepticism $r^{\pm}$ values recorded for reactions known to involve a pre-equilibrium followed by a rate-limiting step. Unfortunately there are many reactions for which the situation is not clear. Thus while there is compelling evidence for a rate-*determining* first step in many electrophilic aromatic substitutions involving displacement of hydrogen (absence of a primary isotope effect) the situation is not clear for many electrophilic reactions involving the displacement of other groups. This may well include some of the many displacements of MR_3 to which the Yukawa—Tsuno equation has been applied (see discussion above).

3.2.3. OTHER COMPLICATIONS

This discussion of the problems posed by a two-step mechanism prompts a more general consideration of the impact of mechanistic complexity on the Hammett equation. The Yukawa—Tsuno equation is basically concerned with detecting and expressing deviations from the appropriate simple LFER, due to variation in cross-conjugation. This, however, is only one of the factors that can produce deviations from simple LFER. Exner has discussed these in detail.[16] Change of mechanism or of rate-controlling step can give Hammett plots showing minima or maxima, or (less extremely) curved plots without turning points. For the latter the deviations from linearity may possibly be removed by use of the Yukawa—Tsuno equation, but $r^{\pm}$ will not be of direct physical significance. Thus various solvolyses show pronounced deviations from conformity to σ^+ in a direction corresponding to $r^+ >> 1.0$, *e.g.* solvolyses of substituted benzyl chlorides. However, this is considered to be due to changing proportions of S_N1 and S_N2 mechanisms, and the r^+ value, if calculated, does not give information on the resonance stabilization of the benzyl cation compared with the t-cumyl cation. In respect of this kind of situation Ruasse and Dubois[17] aptly comment "Such multiparameter correlations can be very misleading if the apparent ρ and r^+ values are not examined critically."

It is difficult to be confident of $r^\pm$ values that are found to be <0.3 or 1 ± 0.3, since reactions characterized by such values will be correlated fairly well by σ (or σ^o) or $\sigma^\pm$ respectively. The procedures involved in calculating $r^\pm$ are such that $r^\pm$ values are not likely to be known with great precision but error estimates are rarely given. This situation is aggravated by there being relatively few common substituents for which $\Delta\sigma_R^+$ is both substantial and reliable; the situation is better for $\Delta\sigma_R^-$, but reactions requiring this tend to be less investigated than those requiring $\Delta\sigma_R^+$. The problem with the substituents of considerable $-R$ effect, and therefore of substantial $\Delta\sigma_R^+$, is that they are subject to interference from such Lewis bases, or protic or Lewis acids as may be present in the role of reactant or solvent. Sometimes this is obviously happening, for dramatic irregularities occur, *e.g.* OMe being apparently less electron-releasing than Me, but more insidious are the small deviations which can mimic behaviour requiring the Yukawa—Tsuno equation.

Applications of the Yukawa—Tsuno equation quite often involve only one or two substituents having considerable $\Delta\sigma_R^+$ values. This may be due to the difficulty of measuring the high rate constants for the compounds with the more highly electron-releasing substituents or to problems of synthesis. Hammett's warning seems timely:[19]

"It does not always seem to be recognized that effective evidence for the Yukawa—Tsuno hypothesis is obtainable only from correlations that include more than one strong through resonating substituent, and for which the value of $r^\pm$ is neither close to unity nor close to zero."

3.2.4. ELECTROPHILIC AROMATIC SUBSTITUTION

Electrophilic aromatic substitution seems to present peculiar difficulties in relation to the Yukawa—Tsuno equation, as was first pointed out by Stock and Brown in 1963.[20] They appreciated the theoretical aims of the Yukawa—Tsuno equation and the reality of at least some of the phenomena with which it had been designed to deal, but they wrote "It is not any easy matter to assess the real merits of the Yukawa—Tsuno correlations .... " and 18 years later this still seems to be true. For electrophilic side-chain reactions they conceded that the equation at least yielded significant improvements in correlations. They were not convinced of this for substitution on the ring and believed that within the limits of precision of the data and other possible sources of discrepancies, simple treatment through σ^+ was successful.

A few years ago Johnson[21] raised again some of the issues with which Stock and Brown dealt. These particularly concerned the 'extended selectivity treatment',[20] a procedure which is based on σ^+ values and deals with the behaviour of a given substituent in several reactions. The logarithms of the partial rate factors for that substituent in a series of electrophilic substitutions are plotted against the corresponding ρ values. One example considered by Johnson is substitution at the β position of thiophen, which is regarded as benzene in which the 'substituent' is S replacing CH=CH. A fairly good straight line is obtained of slope -0.52, which is considered to be the σ^+ value for S replacing CH=CH. The rectilinear relationship is taken to indicate a constancy of resonance effect, irrespective of the electron-demand in the transition state as measured by ρ. Johnson dealt also with other examples, including substitution in the *para-* position of methoxybenzene. He concluded that " ... variation of r^+ from unity is largely due to experimental error."

It should, however, be noted that there is considerable scatter of points about the extended selectivity lines and presumably this too would be attributed to experimental error. Further, the ρ values on which these extended selectivity plots are based are presumably calculated *on the assumption that* σ^+ *correlation is applicable*. The argument is therefore somewhat circular. Thus the ρ value for bromination in aqueous acetic acid at 25°C is taken as -12.1. The

Yukawa—Tsuno equation with the same data gives -9.49. For chlorination the corresponding values are -10.0 and -7.79. Examination of Stock and Brown's[20] plots and text, however, reveals that certain points, including some for *meta*-substituents, are regarded, for various reasons, as not very reliable. Less weight has apparently been given to these in drawing straight lines by eye. Least squares treatment involving the Yukawa—Tsuno equation regards all points as equally reliable. This is a good illustration of the difficulties with which this area is fraught.

Johnson[21] gives one example (from Stock and Brown) which is considered to show a genuine variation of cross-conjugation. This is the extended selectivity treatment of biphenyl, for which the plot is a curve. The variable conjugative effect is attributed to twisting one of the phenyl rings out of the plane of the other: the more negative the value of ρ, the more the rings adopt a coplanar conformation, with an increase in resonance effect across the rings. The variable resonance effect disappears when the two rings are held coplanar, as in fluorene.

3.2.5. A RECENT EXCHANGE OF VIEWS

Johnson[22] has recently criticized the application of the Yukawa—Tsuno equation in another area: S_N1 solvolysis of tertiary *p*-nitrobenzoates of general formula *p*-$NO_2C_6H_4CO_2R$, where R contains an aryl group attached to a tertiary carbinyl centre. An extended selectivity plot for *p*-OMe in about twenty systems is of slope -0.81, almost exactly the value of σ^+ for *p*-OMe (-0.78). It is accordingly claimed that simple correlation analysis involving σ^+ is satisfactory for such systems and the application of the Yukawa—Tsuno equation is meaningless. An explanation is devised to answer the question "Why should the methoxy substituent, and apparently other resonance donors, yield an invariant σ^+ value rather than the spectrum of values predicted?"

There have been sharp rejoinders, however, to this criticism of the Yukawa—Tsuno equation. Lomas and Argile[23] believe "it to be simply an example of circular reasoning" owing to the prominence of the *p*-OMe substituent in the data on which the ρ values for the systems in question are based. These authors support the use of the Yukawa—Tsuno equation with plenty of data. R. Taylor[24] makes very much the same main point in the words "the parameter which the analysis was intended to derive was incorporated in the original data". He examines the situation in great detail and concludes that the shortcomings of the data preclude their use as evidence in favour of the simple σ^+ approach and against the Yukawa—Tsuno equation. More experimental work on well-chosen substrates and substituents appears to be needed, and we may recall Hammett's warning quoted earlier.[19]

At the end of the day one is indeed left with the feeling that there is certainly more work to be done on the Yukawa—Tsuno equation. It seems an inherently reasonable way of expressing through multiple regression the inherently reasonable idea of variability of cross-conjugation and yet the real progress made with it in 20 years does not seem clearly delineated.

Other recent work with a bearing on the Yukawa—Tsuno equation will be referred to after our discussions of Taft's dual substituent-parameter equation and the Swain—Lupton treatment.

3.3. TAFT'S DUAL SUBSTITUENT-PARAMETER EQUATION

Hammett σ values measure the resultant of inductive and resonance effects (see Appendix). Taft and Lewis[25,26] (1958-1959, but see also Taft,[27] 1956) suggested that the resultant effects should be quantitatively separable into inductive and resonance contributions through equations (3.10) and (3.11):

$$\sigma_m = \sigma_I + \alpha\sigma_R \qquad (3.10)$$

$$\sigma_p = \sigma_I + \sigma_R \qquad (3.11)$$

The inductive effect, given by σ_I, is assumed to operate equally from the *meta*- and the *para*-position. The resonance effect, given by σ_R, contributes to σ_m indirectly, α being the 'relay coefficient'.

Taft and Lewis constructed a σ_I scale based on alicyclic and aliphatic reactivities. (See Chapter 4.) For ordinary Hammett σ values, based on the ionization of benzoic acids, a value of α of 0.33 was suggested. [Selected values of σ_I and σ_R are in Table 3.5. To distinguish these σ_R values from those referred to below, and in accord with modern practice, we use the symbol σ_R(BA), *i.e.* based on benzoic acid ionization.] The authors agreed with van Bekkum *et al.*,[11] however, in regarding the ionization of benzoic acids as a not entirely satisfactory standard process (see Section 3.1.1). They proposed the use of 'insulated' reaction series, *i.e.* reactions in which the reaction centre is incapable of conjugation with the ring at any stage. The ionization of phenylacetic acids is such a series. Sigma values based on such processes were designated σ^o, and they should correspond to the σ^n of van Bekkum *et al.*,[11] although derived by a different procedure. Selected values of σ^o are shown in Table 3.5. The analysis into inductive and resonance effects may also be performed for σ^o constants, giving σ_R^o values (α = 0.5) or with σ^+ and σ^- constants giving σ_R^+ and σ_R^- respectively. (See Table 3.5 for a selection; also Exner[4] for extensive compilations.)

TABLE 3.5. Selected values[a] of σ_I, σ_R, σ^o, σ_R^o, σ_R^+, and σ_R^-

Substituent	σ_I	σ_R(BA)	σ_m^o	σ_p^o	σ_R^o	σ_R^+	σ_R^-
Me	-0.05	-0.11	-0.06	-0.14	-0.10	-0.25	-
COMe	0.28	0.20	0.36	0.47	0.16	-	0.56
NH_2	0.19	-0.79	-0.09	-0.30	-0.48	-1.61	-
NO_2	0.65	0.15	0.71	0.81	0.15	-	0.6
OH	0.25	-0.63	0.02	-0.20	-0.43	-1.2	-
OMe	0.27	-0.61	0.10	-0.14	-0.45	-1.02	-
F	0.50	-0.45	0.34	0.15	-0.34	-0.57	-
Cl	0.46	-0.23	0.37	0.24	-0.23	-0.36	-
Br	0.44	-0.20	0.37	0.26	-0.19	-0.30	-
I	0.39	-0.17	0.34	0.28	-0.14	-0.25	-

a σ^o values are from the compilation by Exner.[4] σ_I and the σ_R-type parameter values are mainly from Ehrenson, Brownlee and Taft,[29] sometimes with slight changes to make the values more consistent with those of other σ values in the Tables of the present Chapter. It must again be emphasized that Tables of σ values in the present book must not be used uncritically.

The importance of the separation of sigma parameters into σ_I and σ_R-type contributions is that it suggests the possibility of a 'dual substituent-parameter' treatment for reaction series through an equation of the form (3.12):[28]

$$\log (k/k^o) = \rho_I \sigma_I + \rho_R \sigma_R \qquad (3.12)$$

The simplest situation for applying this equation would arise if the various σ_R-type scales distinguished above turn out to be highly collinear. It should then be satisfactory to characterize each substituent by σ_I and, say, σ_R^o and apply equation (3.12) to *meta* and *para* reaction series separately; *i.e.* with the Hammett equation we have, for each substituent, position-dependent sigma values σ_m and σ_p (arbitrarily becoming σ^+ or σ^- on occasion) but a single ρ value for each reaction series, while in Taft's treatment each substituent is characterized by position-independent σ_I and σ_R^o values, but the susceptibility to inductive and to resonance effects is to be expressed separately through position-dependent ρ_I and ρ_R values.

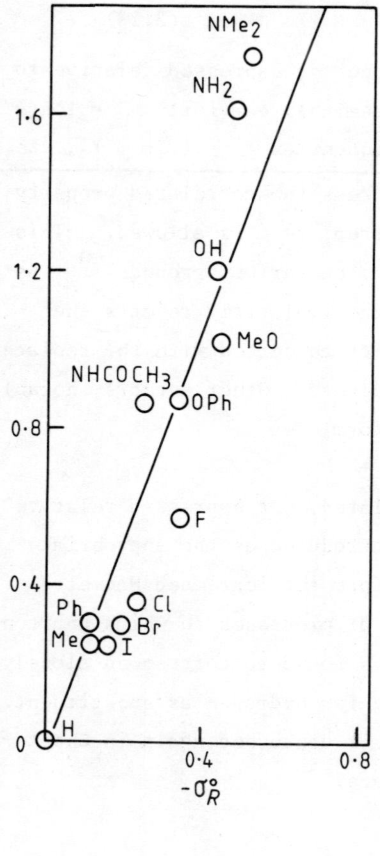

FIG. 3.1. σ_R^+ *versus* σ_R^o for selected substituents. Values mainly from Ehrenson, *et al.*[29]

Thus the dual substituent-parameter (DSP) equation is from the start a considerably more complicated system of correlation analysis than the Hammett equation (as above) or the Yukawa— Tsuno equation. As will be seen shortly, its complications are, in fact, greater than indicated above, because the hoped-for collinearity between the various σ_R-type scales is not sufficient for the really precise correlation analysis of a wide range of reactions and other processes, *i.e.* the σ_R-type values for the various substituents change relatively to each other when the electron-demand is varied (*cf.* Fig. 3.1).

The historical development of the DSP equation has recently been reviewed in detail by the present author.[15] In this Section only a few main topics can be mentioned, and we will concentrate on the current position.

Taft and his colleagues now write the equation in the form:[29]

$$P^i = \rho_I^i \sigma_I + \rho_R^i \sigma_R \qquad (3.13)$$

P^i is the substituent-influenced property, expressed relative to hydrogen as substituent, *e.g.* for chemical reactivity $P^i = \log(k/k^{\circ})_i$ or $\log(K/K^{\circ})_i$, for spectroscopic frequencies $P^i = (\nu - \nu^{\circ})_i$, *etc.*†

† Taft and his colleagues always express the correlated property relative to that for H, and no intercept term is allowed. This procedure is justified by the authors on various grounds.[30] As noted in Section 3.1, such a procedure implicitly rejects the possibility of there being some effect connected with the replacement of H by any other substituent whatsoever. Other authors, notably Charton,[31] use the equation in the form

$$Q = \alpha\sigma_{I,X} + \beta\sigma_{R,X} + h$$

where Q is the property to be correlated, *not* expressed relative to hydrogen as substituent, and h is introduced as the appropriate intercept term. We may call this form the 'extended Hammett equation'. In the enormous variety of processes to which Charton has applied this equation, h is not always found to correspond closely to the value of the correlated property for hydrogen as substituent. The question of "intercept or not?" is discussed again in Chapter 7, along with various statistical matters.

The right-hand side of the equation attributes the effect of substituents in ring position i to a linear combination of a polar (inductive) effect ($I^i = \rho_I^i\sigma_I$) and a resonance effect ($R^i = \rho_R^i\sigma_R$). The susceptibility constants ρ_I^i and ρ_R^i define the blend of polar and resonance effects characteristic of the influence of substituents in the position i on the property in question under given conditions, the blending constant λ being defined as ρ_R^i/ρ_I^i.‡ As already noted, the substituent constants σ_I and σ_R are position-independent and the Taft equation is applied separately to the effects of *meta*- and *para*-substituents.

The σ_I scale has long been held to be of wide general applicability but the possibility of defining a unique, widely applicable σ_R scale is an important issue and has been much discussed. As early as 1959 Taft[26] stated that a single, widely applicable, and precise σ_R scale could not be devised. In spite of this, much correlation analysis has been done on the assumption that a single σ_R scale *is* widely applicable. Thus, Wells, Ehrenson, and Taft[32] used σ_R^o values in their analysis of naphthalene reactivities, although the reaction series involved might be expected to show considerable variation in the importance of resonance effects. Further most of Charton's work (see footnote † above) has involved the σ_R(BA) scale based on benzoic acid ionization through $\sigma_R(\text{BA}) = \sigma_p - \sigma_I$. The matter was thus in a somewhat indecisive state for a long time, but considerable clarification was provided by the work of Ehrenson, Brownlee, and Taft,[29] which compared thoroughly the success of various σ_R scales in the application of equation (3.13) to a wide variety of processes. We must now summarize the salient features of this work.

The initial aim was to define satisfactorily four scales of σ_R values: σ_R^o, σ_R(BA), σ_R^-, and σ_R^+. In each case this was done by applying least-squares fitting procedures to data for selected reaction series expected to exhibit resonance effects of the required

‡ The question of assessing the relative importance of 'effects' in multiple regression will be given a more detailed discussion in Chapter 7.

type. There is great stress on the requirement that in each series data shall be available for a *minimal basis set of substituents* as follows: NMe_2, NH_2, or OMe (any two); any two $+R$ substituents (CF_3, CO_2R, CH_3CO, CN, NO_2): H and CH_3 (both); two halogens (but not both Cl and Br). This is held to be necessary to minimize collinearity between inductive and resonance parameters. The article should be consulted for the wealth of detail involved in setting up the various scales. Twenty-four substituents (including H) were considered. A fixed σ_I scale (based on a careful reassessment of alicyclic and aliphatic reactivities) was used throughout. The σ_R(BA) scale was based on eight reaction series (mainly ionizations of benzoic acids in various solvents). Only for the ionization of benzoic acids in water at 25°C was the condition $\rho_I^p = \rho_R^p$ imposed (to establish the scale).† The σ_R^- scale was based on the ionization of anilinium ions in water at 25°C. (It is interesting that the ionization of phenols in water was found unsatisfactory for this purpose.) The σ_R^o scale was based on 14 'insulated' reaction series (mainly $ArCH_2X$ reactivities) and also on three ^{19}F nmr sets.‡

† This scale-establishing condition for the ionization of benzoic acids in water at 25°C appears to be the sole remnant of the assumptions, originally widely applied, about the inductive effect operating equally from the *meta*- and the *para*- position and the variable relay factor for the operation of the resonance effect from the *meta*-position. These are now matters for investigation through the ρ_I^m, ρ_I^p, ρ_R^m, and ρ_R^p of the 'unconstrained' equation (3.13).

‡ The correlation analysis of ^{19}F nmr has played a key role in the development of the DSP equation (see review by Shorter[15]). It was this that first indicated that the susceptibilities of a property to inductive and resonance effects need not be the same, even though they are so defined in the reference process, that is benzoic acid ionization ($\sigma_p = \sigma_I + \sigma_R$), and appeared also to be the same in a variety of processes.

The σ_R^+ scale was based on a combined consideration of t-cumyl chloride solvolysis and protodesilylation of phenyltrimethylsilanes (Eaborn[33]). (The t-cumyl chloride reaction does not meet the minimal basis set requirement.)

The application of σ_I and each of the σ_R scales in turn was examined for a large number of series for reactivity (equilibrium and rates), ir, and nmr data of *para*-substituted compounds. The series were classified under headings relating them to the various σ_R scales. It was shown very strikingly that for any given series the best correlation† is nearly always given by the σ_R scale based on the process which obviously bears the closest relationship to the series in question, the other σ_R scales being markedly inferior. The authors concluded that each of the σ_R scales is necessary for a satisfactory comprehensive system of correlation analysis for substituent effects in benzene, each scale having a 'limited generality'.

3.3.1. THE INTERPRETATION OF ρ_I AND ρ_R VALUES

The interpretation of ρ in the original Hammett equation is well developed and there is good qualitative and some quantitative understanding of how it is governed by factors such as reaction type, solvent, temperature, structural characteristics, *etc.*[6,8] Corresponding distinctive interpretation of ρ_I and ρ_R, and the ratio $\rho_R/\rho_I = \lambda$, is not yet so well developed, but some progress has been made, which shows the deeper insights that may be provided by the DSP equation, compared with the original Hammett equation. We restrict ourselves to a few examples drawn from the work of Ehrenson, *et al.*[29]

A comparison of λ^m and λ^p values for the same reaction clearly shows the poorer transmission of resonance effects from the *meta*-

† Taft and his colleagues use f = SD/RMS as a measure of the success of correlations. SD is the root mean square of the deviations and RMS is the root mean square of the data P_i's. If the value of $f < 0.1$, the correlation is considered of good precision. The status of this f statistic is discussed in Chapter 7.

position. While λ^m/λ^p is typically *ca.* 0.4, there is considerable reaction dependence, which will no doubt receive detailed interpretation in due course.

For the ionization of benzoic acids ρ_R^p and ρ_I^p both increase with decreasing solvent polarity, but λ^p decreases at the same time. The latter is attributed to a decreasing relative importance of cross-conjugation as the polarity of the solvent is decreased. The interpretation is confirmed by the absence of a solvent effect on λ^p for phenylacetic acids and other 'insulated' systems.

Moreover, ρ_I and ρ_R may be of opposite sign so that λ is negative, *e.g.* for the rates of acid-catalysed esterification of benzoic acid in methanol, 25°C, ρ_I^p is negative because passage from initial state to transition state involves development of positive charge on the side-chain. ρ_R^p is positive because there is parallel loss of cross-conjugative stabilization of the initial state by $-R$ substituents.†

Extending a conjugated system of aromatic ring and reaction centre by incorporating C=C or C≡C appropriately in the side-chain has little effect on λ^p. However, steric twisting of the ring and reaction centre from coplanarity greatly reduces λ^p, indicating inhibition of transmission of resonance effects.

3.3.2. THE SCALING OF σ_I AND σ_R VALUES

In the early 1960s there was considerable discussion over what then appeared to be a basic assumption of Taft's separation of inductive and resonance effects, *i.e.* the supposed equality of inductive effects as between the *meta*- and the *para*-position. As a

† It seems likely that equation (3.13) is not liable to some of the troubles of the Yukawa— Tsuno equation in respect of the interpretation of the reaction constants. Thus for two-step processes for which $k = Kk_2$, the observed ρ_I and ρ_R values will be, respectively, the sums of the ρ_I and ρ_R values of the individual steps, a simple significance compared with that of the Yukawa— Tsuno $r^{\pm}$ value in the same situation (equation 3.9). These considerations permit the simple interpretation for acid-catalysed esterification given above.

result of studies of the dissociation constants of *meta*- and *para*-substituted benzoic acids involving substituents incapable of ordinary conjugative effects, Exner[34] concluded that the inductive effect was stronger from the *para*- than from the *meta*-position by a factor of 1.14. He further adduced evidence that the usual σ_I values based on alicyclic and aliphatic reactivities were out of scale with σ_m and σ_p by a factor of 1.10, and should be multiplied by this factor to introduce the π-inductive component which would be absent in the saturated systems. These considerations led to a revised analysis of σ_m and σ_p values in terms of revised scales of σ_I and σ_R. (Selected examples are in Table 3.6.)

TABLE 3.6. Selected values of σ_I and σ_R derived from Exner's analysis[34]

Substituent	σ_I [a]	σ_R [b]
Me	-0.06	-0.07
COMe	0.34	0.05
NH_2	0.11	-0.69
NO_2	0.70	0
OH	0.28	-0.68
OMe	0.31	-0.63
F	0.56	-0.59
Cl	0.51	-0.35
Br	0.50	-0.34
I	0.43	-0.28

a Values obtained by bringing aliphatic or alicyclic values to the same scale as σ_m and σ_p by multiplying by 1.10.
b Values obtained by subtracting from σ_p the scaled values of σ_I multiplied by 1.14. The values of σ_I and σ_R have been recalculated for this Table and in some cases differ slightly from those given by Exner.[34]

The status of Exner's revised σ_I and σ_R values has not altogether been settled. Their most surprising feature is that $\sigma_R = 0$ for certain $+R$ substituents such as NO_2 and CN, a finding that most physical organic chemists have found hard to accept (see discussion by Shorter[15]). Taft and his colleagues[32] have examined the application of Exner's scales and have concluded that they do not have any marked advantage over the usual, more simply based σ_I and σ_R scales. However, in the derivation of, say, σ_R^o values by subtraction of $\gamma\sigma_I$ from σ_p^o, it is difficult to eliminate the possibility that γ should differ from unity by up to $\pm$ 0.2.

Recently Charton[35] has advocated the multiplying of the normal σ_I values by 1.09 before using them to calculate σ_R values. By a consideration of model systems he has also found support for Exner's contention that the inductive effect operates more powerfully from the *para*- than from the *meta*-position. Charton has also advocated replacing the symbols σ_I and σ_R by σ_L and σ_D. The subscripts *L* and *D* denote 'localized' and 'delocalized' respectively, and these words are held to be more appropriate descriptions of the electronic effects than the customary terms 'inductive' and 'resonance'. Exner[36] has himself continued to advocate his views on this question. It is possible that part at least of what is at issue here is the status of the π-inductive effect. Is it to be regarded primarily as an *inductive* effect and put into σ_I, or as a *resonance* effect and put into σ_R? It is doubtful whether one of these procedures is 'right' and the other 'wrong'. It is a matter of the choice of models for the analysis of the total substituent effect.

3.4. THE SWAIN-LUPTON TREATMENT

We can deal with this only rather briefly. For a fuller review and appraisal, see Shorter.[15]

Swain and Lupton[10] strongly criticized the proliferation of substituent constants and maintained that any one of the various σ scales† was expressible as a linear combination of any other two, *i.e.*

$$\sigma_3 = a\sigma_1 + b\sigma_2 \qquad (3.14)$$

They illustrated this point by correlating various σ scales with σ_m and σ_p based on benzoic acid ionization. This was actually done by the use of

$$\sigma = a\sigma_m + b\sigma_p + i \qquad (3.15)$$

† In this section σ is often used in the general sense of polar substituent constant, and not just for the benzoic acid scale.

to avoid forcing the regression through the origin, "since measurements on the unsubstituted compound are in general no more accurate than those on substituted ones". However i was in general indistinguishable from zero. The authors were satisfied with most of the correlations presented and commented favourably on the high correlation coefficients.† Thus Taft's σ_m^o and σ_p^o values are well correlated (in separate equations); the correlations are less, but still acceptably, successful for σ_p^+ and σ_p^-. Swain and Lupton thus maintained that all correlation analysis for effects of *meta*- and *para*- substituents could be done by multiple regression on σ_m and σ_p, but a deficiency in this procedure would lie in the hybrid nature of σ_m and σ_p, both being blends of field‡ and resonance effects. The authors thus sought to establish numerical scales for just two basic electronic characteristics of substituents: a field‡ constant $\mathfrak{F}$ and a resonance constant $\mathfrak{R}$. Two assumptions were necessary to enable this to be done.

The first assumption was that the effect of 4-substituents on the dissociation constant of bicyclo[2.2.2]octane-1-carboxylic acid is entirely a field effect, because there is no intervening conjugation or unsaturation. $\mathfrak{F}$ values are therefore based on the $\Delta \log K_a$ values of these acids. It appears to have been the authors' intention to divide these $\Delta \log K_a$ values by the corresponding ρ value for the ionization of benzoic acids under the same conditions, thereby identifying the scale with the σ' scale of Roberts and Moreland[37] (see Chapter 4). However, the raw $\Delta \log K_a$ values were actually used (further discussions later). For 14 substituents (the only ones for which there were data from the bicyclo-octane system) the $\Delta \log K_a$ values were correlated with σ_m and σ_p and the

† Swain and Lupton do not use multiple correlation coefficient as ordinarily defined but a 'coefficient of determination'. The significance of this quantity is discussed in Chapter 7.

‡ This term as used by Swain and Lupton includes the 'pure field (direct) effect and the inductive (through-bond) effect. It is thus synonymous with the loose usage of 'inductive' (see Appendix).

resulting expression was then used to calculate a self-consistent set of $\mathcal{F}$ values for 42 substituents.

The $\mathcal{R}$ scale was essentially based on applying a general equation relating σ values with $\mathcal{F}$ and $\mathcal{R}$:

$$\sigma = f\mathcal{F} + r\mathcal{R} \tag{3.16}$$

to σ_p in the form

$$\sigma_p = \alpha\mathcal{F} + \mathcal{R} \tag{3.17}$$

Thus the field f and resonance r weighting factors were taken as α and 1.00 respectively for σ_p. The second assumption, necessary to establish the value of α, was that for $\overset{+}{N}Me_3$, $\mathcal{R} = 0$. It was argued that resonance interaction should be impossible for this substituent, because the nitrogen atom in the quaternary state can neither accept nor donate electrons. This was held to be supported by the negligible effect of $\overset{+}{N}Me_3$ on the uv spectrum of benzene. In equation (3.17) σ_p, $\mathcal{F}$, and $\mathcal{R}$ were put equal to +0.82, +1.46, and 0 respectively for $\overset{+}{N}Me_3$, and hence α was calculated to be 0.56. $\mathcal{R}$ values could then be calculated from σ_p and $\mathcal{F}$ values for the substituents considered. Values of $\mathcal{F}$ and $\mathcal{R}$ for selected substituents are in Table 3.7.

Swain and Lupton claimed that the interrelationships of the $\mathcal{F}$ and $\mathcal{R}$ values for the 42 substituents considered show that the parameters are "reasonable" measures of field and resonance effects. The $\mathcal{F}$ and $\mathcal{R}$ sets were used by Swain and Lupton as new independent variables in multiple regression, in support of their contention that the various σ scales may be regarded as linear combinations of $\mathcal{F}$ and $\mathcal{R}$. σ_m and σ_p are of course perfectly reproduced since they are involved in the setting up of the $\mathcal{F}$ and $\mathcal{R}$ scales. σ^o_m and σ^o_p are also well correlated, but the performance for σ^+_p and σ^-_p is not so good (*cf.* above).

The authors also assessed the 'percent resonance' character of the various series by means of equation (3.18), where the vertical lines

$$\%\mathcal{R} = \frac{100\ \psi|r|}{\phi|f| + \psi|r|} \tag{3.18}$$

TABLE 3.7. Swain–Lupton Field and Resonance Parameters

Substituent	As calculated by Swain and Lupton[10] $\mathcal{F}$	$\mathcal{R}$	As modified by Hansch *et al.*[38] $\mathcal{F}$	$\mathcal{R}$
NH_2	0.037	-0.681	0.02	-0.68
OCH_3	0.413	-0.500	0.26	-0.51
SCH_3	0.332	-0.186	0.20	-0.18
CH_3	-0.052	-0.141	-0.04	-0.13
C_6H_5	0.139	-0.088	0.08	-0.08
F	0.708	-0.336	0.43	-0.34
Cl	0.690	-0.161	0.41	-0.15
Br	0.727	-0.176	0.44	-0.17
I	0.672	-0.197	0.40	-0.19
H	0.000	0.000	0.00	0.00
CF_3	0.631	0.186	0.38	0.19
CO_2R	0.552	0.140	0.33	0.15
CN	0.847	0.184	0.51	0.19
NO_2	1.109	0.155	0.67	0.16

denote absolute magnitudes, and ϕ and ψ are scale factors to correct for the different numerical ranges of the $\mathcal{F}$ and $\mathcal{R}$ scales, and are defined as in equations (3.19) and (3.20), the number of substituents considered being 42. $\%\mathcal{F}$ is 100 - $\%\mathcal{R}$. The values of $\%\mathcal{R}$ are usually

$$\phi = \frac{\Sigma|\mathcal{F} - \bar{\mathcal{F}}|}{42} = 0.365 \qquad (3.19)$$

$$\psi = \frac{\Sigma|\mathcal{R} - \bar{\mathcal{R}}|}{42} = 0.228 \qquad (3.20)$$

qualitatively reasonable, *e.g.* $\%\mathcal{R}$ is greater for σ_p than for σ_p^o, and is greater for σ_p^+ than for σ_p.

Swain and Lupton concluded that $\mathcal{F}$ and $\mathcal{R}$ are successful overall in accounting for the various σ scales and that the introduction of additional parameters would not be significant. In their view "it appears practicable to drop the use of about 20 other synthetic hybrid sets, which are less pure field or resonance measures".

Later Hansch *et al.*[38] pointed out that Swain and Lupton omitted to divide the Δ log K_a values of the 4-X-bicyclo[2.2.2.]octane-1-carboxylic acids, by the ρ value for the ionization of benzoic acids under the same conditions (50% w/w ethanol-water, 25°C). Taking ρ = 1.65 (the basis for this is not clear, but the value at first sight is not unreasonable; see, however, Shorter[15]), Hansch suggested that Swain and Lupton's $\mathfrak{F}$ values are out of scale with $\mathfrak{R}$ by a factor of 1.65. Hansch *et al.* therefore reworked and greatly extended the Swain—Lupton procedure to calculate $\mathfrak{F}$ and $\mathfrak{R}$ values for a large number of substituents, initially 191 but very recently extended to many more.[5] (Examples are in Table 3.7.)

3.4.1. APPLICATIONS OF THE SWAIN—LUPTON TREATMENT

Since 1968 the treatment has been widely applied in the same general way as the DSP equation, *i.e.* to *meta*- and *para*-substituted reaction series separately, and its performance has often been compared with that of the DSP equation. Shorter[15] has recently reviewed some of the applications. Here we can give only a very brief account.

The treatment has undoubtedly found most application in the correlation analysis of nmr data, see Ewing.[39] Workers in the field of nmr seem more attracted to multiparameter treatments than are physical organic chemists in general, because of the undoubted difficulty of obtaining satisfactory correlations of nmr data with any one of the individual scales of σ values. While some workers in nmr are enthusiastic about $\mathfrak{F}$ and $\mathfrak{R}$, others are less so and believe the DSP equation to be superior. There are also applications in optical spectroscopy.

$\mathfrak{F}$ and $\mathfrak{R}$ are held by some to be peculiarly suitable for the correlation analysis of biological activity data. In the correlation analysis of a given biological activity for a series of compounds, it is often impossible to classify substituents as *ortho*, *meta*, or *para* with respect to a certain point in the molecule believed responsible for the biological activity. The position-independent parameters $\mathfrak{F}$ and $\mathfrak{R}$ have thus an obvious advantage over the position-dependent σ_m and σ_p. Further, as pointed out by Hansch *et al.*,[38]

$\mathfrak{F}$ and $\mathfrak{R}$ values for a series of substituents tend to vary in a mutually orthogonal manner, whereas σ_m and σ_p are often fairly highly correlated. $\mathfrak{F}$ and $\mathfrak{R}$ are therefore much more suitable than σ_m or σ_p for inclusion with other structural parameters in stepwise multiple regression and other regression procedures (see Hansch[40]).

Applications to chemical reactivity seem restricted to a few instances in some of the main areas, *e.g.* acidic and basic strength, electrophilic or nucleophilic substitution (ring or side-chain), and electrophilic addition. Here also the performance is mixed, with some claims for the superiority of the Swain—Lupton treatment over the DSP equation, and *vice versa*. There is perhaps a tendency for the Swain—Lupton treatment to become less satisfactory when σ^+ would be relevant in a one-parameter treatment.

There is also a treatment of the *ortho*-effect based on the Swain—Lupton treatment, to which reference will be made in Chapter 4.

3.4.2. CRITIQUE OF THE SWAIN—LUPTON TREATMENT

The treatment has won some success as an empirical correlation procedure, although it is occasionally stated to be inferior to other types of multiparameter treatment or to accomplish little more than a one-parameter treatment.

Ehrenson *et al.*[29] considered the Swain—Lupton treatment in connection with their examination of the need for various σ_R-type scales (Section 3.3). Only rarely is the Swain—Lupton treatment superior to that involving the best σ_R scale, and often it is vastly inferior. The Swain—Lupton approach is particularly poor where σ_R^+ or σ_R^- is best. The authors are very critical of Swain and Lupton's exclusion of important substituents such as NMe_2, and of any reactivities of the σ^- type except the ionization of phenols in water (a process shown to have its own peculiarities). They conclude that "there appears to be little to encourage proliferation of these parameters".

It is certainly significant that $\mathfrak{F}$ and $\mathfrak{R}$ are rather less successful in expressing σ^+ or σ^- scales than they are in dealing with σ scales involving less extreme resonance effects. Exner[16] pointed out that the reproduction of σ^+ and σ^- with a standard deviation of as large as 0.18 cannot be considered satisfactory. Many of Swain and Lupton's "revised" σ^+ and σ^- values (*i.e.* those given by the expressions for the regression of σ^+ or σ^- on $\mathfrak{F}$ and $\mathfrak{R}$) are unacceptably far from the experimental values based on the defining reactions, *e.g.*

F, σ_p^+ (SL) = −0.247, σ_p^+ (obs) = −0.07;
OMe, σ_p^+ (SL) = −0.648, σ_p^+ (obs) = −0.778;
SMe, σ_p^+ (SL) = −0.164, σ_p^+ (obs) = −0.604.

The Swain—Lupton treatment may be justly criticized, however, not on the grounds that it does not work as a system of correlation analysis (it works as well as any system with a fixed scale of resonance effects) but that it claims to make a separation of substituent effects that is illusory.

It is the values of $\mathfrak{R}$ rather than those of $\mathfrak{F}$ that are criticized. ($\mathfrak{F}$ values are basically related to σ_I values through the use of the bicyclo-octane model.) Several authors[41] have pointed out that $\mathfrak{R}$ values depend critically on the assumption that NMe_3^+ has no resonance effect.

This group is, however, isoelectronic with Bu^t, which is known to be a $-R$ group, and there seems to be no reason for excluding *a priori* this possibility for NMe_3^+. Evidence of the $-R$ character of NMe_3^+ comes *inter alia* from the ir spectra of *para*- and *meta*-disubstituted benzenes, from ^{19}F nmr study of 4-fluorotrimethylanilinium ion and 4-fluoro-4'-trimethylammonio-*trans*-stilbene, from the σ_p^+ value of NMe_3^+, and from pK_a values.[41] In any case it may be questioned whether it is wise to base a treatment of the effects of substituents, which are for the most part dipolar, on the unipole NMe_3^+. There is the further point that $\sigma_p^+ = +0.82$ for NMe_3^+ is given an uncertainty of $\pm >0.2$ by McDaniel and Brown.[42] The extent to which the treatment in fact separates field (inductive) and resonance effects must be considered doubtful.

3.4.3. EXTENSIONS OF THE SWAIN—LUPTON TREATMENT

It should be mentioned that Hansen and Hepler[43] have a treatment of substituent effects that is said to be an extension of that of Swain and Lupton. Also S. H. Unger[44] has developed a treatment of substituent effects which is related to that of Swain and Lupton. It has already been mentioned that $\mathfrak{F}$ and $\mathfrak{R}$ tend to vary in a mutually orthogonal manner. Unger seeks to develop substituent constants S and P that possess this property to an even greater extent and thereby accomplish a more complete separation of field and resonance effects. While S is based initially on the ionization of bicyclo-octane carboxylic acids (as is $\mathfrak{F}$) and P values are appropriately guessed, the final values (for 42 substituents) emerge from a computerized procedure that optimizes the account given of the selected body of data. It is claimed that S and P give, on average, higher correlation coefficients than other pairs of parameters, even for data not included in the optimization. For details the thesis must be consulted. A thesis by N.R. Rosenquist is also relevant.[45]

Williams and Norrington[46] carried out a statistical analysis of data from 45 reaction series to determine positional weighting factors f and r for $\mathfrak{F}$ and $\mathfrak{R}$ respectively, in order to permit the treatment of *ortho*-, *meta*-, and *para*-substituted derivatives in the same regression. The values of f_p and r_p were defined as 1.000, f_m was found to be 0.980 (not significantly different from 1.000), f_o 1.248, r_m 0.347, and r_o 0.863. It was claimed that the values "agree roughly with the *ad hoc* assumptions made hitherto by classical organic chemists and enable good correlations to be found between physico-chemical properties of the side-chain and substituent parameters of mixed or multiply-substituted populations of benzenoid compounds". However the inclusion of *ortho*-substituted derivatives often makes the correlations less successful. A recent application of Williams and Norrington's approach is provided by L.N. Patnaik, *et al.*[47] for some autoxidation and hydrogen abstraction reactions.

In conclusion we repeat that the Swain—Lupton treatment may be criticized, not on the grounds that it does not work as a system of correlation analysis (it works as well as any system with a fixed scale of resonance effects), but that it claims to make a separation of substituent effects which is illusory.[15]

3.5. MISCELLANEOUS TOPICS AND SOME RECENT DEVELOPMENTS

In the logical development of this Chapter, there has been no opportunity to mention certain work which involved what were probably the earliest applications of multiple regression in physical organic chemistry. Jaffé[3] discussed the correlation analysis of several types of system which required a multiparametric extension of the Hammett equation of the form:

$$\log k \;=\; \log k^{o} \;+\; \rho_1\sigma_1 \;+\; \rho_2\sigma_2 \qquad (3.21)$$

This equation was examined in relation to the effect of substituents in both rings of a substrate Ar^1Ar^2Y, in which the rings were not disposed symmetrically with respect to the reaction centre Y, and to the effect of substituents in both rings for a reaction of the type:

$$Ar^1Y^1 \;+\; Ar^2Y^2 \;\rightarrow\; \text{products} \qquad (3.22)$$

Jaffé obtained several satisfactory correlations for systems of the former type but not for those of the latter type, although satisfactory results for such systems have been obtained later by other workers. Clearly equation (3.21) requires that the substituent effects in the two rings are essentially independent. Good examples of the two types of system are provided (respectively) by the alkaline hydrolysis of acyl- and aryl-substituted phenyl benzoates[48] $Ar^1CO_2Ar^2$ in 33% acetonitrile-water, 25°C, for which ρ_1 = 2.037, ρ_2 = 1.320, n = 24, R = 0.997, s = 0.092,† and the reaction of $Ar^1_2CN_2$ with Ar^2CO_2H in toluene, 25°C, for which ρ_1 = -1.62, ρ_2 = 2.376, n = 46, R = 0.998, s = 0.078.[49]

When correlation analysis with (3.21) gives poor results, the inclusion of a 'cross-term' involving $\sigma_1\sigma_2$ may be examined.[50,51]

Jaffé[52] later suggested other applications of equation (3.21), including cases in which 1 and 2 denote alternative paths for transmission of electronic effects of substituents to the reaction site. A good example of this situation is provided by the relative

† The data have been reanalysed for the present work. For p-NO_2 in Ar^2 the ordinary σ value of 0.778 was used, *cf.* the special value of 0.89 suggested by the authors.[48]

half-neutralization potentials of substituted salicylic acids in benzene.[53] Here 1 and 2 refer to the position of a substituent relative to the carboxyl group and to the phenolic group respectively, and thus to transmission of the electronic effect either directly or *via* the phenolic hydrogen bond. Good correlation was found with ordinary σ values for σ_1 and σ^- values for σ_2, and ρ_2/ρ_1 was found to be ≈ 0.4, indicating that transmission *via* the hydrogen-bond makes a substantial contribution. For the log K values in water, however, there was little indication of transmission *via* the hydrogen bond.[54]

The account of the Hammett equation and its extensions as presented in this Chapter has been concerned largely with the reactivitiy of *meta*- and *para*-substituted benzene derivatives, *i.e.* the area for which the Hammett equation was originally devised. The equations have, however, been applied in various ways to the behaviour of appropriately substituted polycyclic arenes and heterocyclic systems. The reader is referred to the comprehensive review of these areas by Charton.[31]

An important topic in the correlation analysis of polycyclic systems is the Dewar—Grisdale treatment,[55] which is sometimes regarded as generating multiparameter extensions of the Hammett equation. Indeed the F and M parameters of this treatment are sometimes used in multiple regression as if they were analogous to Taft's σ_I and σ_R-type constants. In our view the Dewar—Grisdale treatment is primarily concerned with the analysis of σ_m and σ_p (as originally defined) into field and mesomeric contributions, on the basis of a definite physical model of polar substituent effects, and the use of this model to predict σ values for the various substituent—reaction-centre dispositions in polycyclic systems such as naphthalene. The treatment is not much concerned with the variability of cross-conjugation, although there is some reference to this in a later paper.[56]

Mention of the Dewar—Grisdale treatment prompts a reference to the whole question of the interpretation of substituent and reaction constants in terms of physical models. We have referred earlier to the qualitative understanding of substituent constants and to the mechanistic interpretation of the reaction constants in the multi-parameter extensions of the Hammett equation, *e.g.* the $r^{\pm}$ of the Yukawa—Tsuno equation, and ρ_I^i and ρ_R^i of the DSP equation. Considerable efforts have also been made, and are currently being made, in the direction of a more quantitative interpretation of substituent and reaction constants. Various types of work are involved. Thus attempts are being made to develop physical models for the transmission of the non-mesomeric part of the electronic effect from substituent to reaction centre, *i.e.* to establish the real nature of the so-called inductive effect as measured by σ_I. Much of this work involves the use of systems designed to distinguish the relative contributions of the field (or direct) effect and through-bond polarization effects (involving σ- or π- electrons). The importance of the field effect is now well established,[16] and during the past decade many authors have considered the role of through-bond polarization to be small. However, more emphasis is now being laid on the role of through-bonds effects in certain systems. Some authors are attempting a very detailed analysis of the nature of electronic effects. The reader is referred especially to papers by Exner[57] and by W. F. Reynolds,[58,59] which contain extensive bibliographies.

Another important area is the interpretation of reaction constants through the application of quantum mechanics, *e.g.* the attempt to understand the relative ρ values for the ionization of CO_2H attached to various carbon skeletons. See, for example, recent papers by Ponec[60] and by Exner,[61] which employ HMO and CNDO approximations with some success. In a recent paper, Forsyth and Sandel[62] apply the DSP equation involving σ_I and σ_R^+ to data for the solvolysis of 1-arylethyl derivatives and discuss the resulting ρ_R^+ values in relation to MO calculations of charge distribution and 'HOMO-LUMO interactions'. However, the most striking relationship discovered was between the ρ_R^+ values and the amount of charge developed at the site of substitution, as inferred from ^{13}C chemical shift data of stable arylcarbenium ions. This also illustrates the current importance of ^{13}C nmr studies in relation to substituent effects.[39,63-71]

We must mention also the important development of experimental techniques for studying in the gas-phase many types of process hitherto studied only in solution and the development of theoretical techniques for treating molecular systems. This has also made possible the analysis of substituent effects into intrinsic molecular effects and solution effects, with consequent clearer understanding of substituent effects. Here we can do no more than give a few recent references for experimental techniques[72,73] and to the use of *ab initio* molecular orbital calculations.[74,75]

The inclusion of data for unipolar substituents, such as NMe_3^+, SMe_2^+, or CO_2^-, in reaction series has long posed problems in correlation analysis. In an article on the electronic effects of sulphonio groups (written in 1978 and published in 1981) the present author drew attention to these and wrote:[76] "it is most unwise to analyse or interpret the substituent constants of these groups, to compare values with those for dipolar or other unipolar groups, or to include data for sulphonio-substituted compounds in correlations, without regard to the limitations which these substituent constants inevitably have". However, some substantial progress towards a proper extension of the Hammett equation to unipolar groups has now been made. The reader is referred to an important paper by Wepster and his colleagues.[77] The essence of the treatment lies in considering a "Bjerrum field effect term" as a separate component of the Hammett equation. (Earlier work of Koppel, Palm, and their colleagues[78] on substituent effects in aliphatic systems had indicated that the solution to the problem of unipolar substituents might lie in the separation of electrostatic interaction from the rest of the substituent effect.)

There are several recent papers concerned with fundamental aspects of the multiparameter extensions of the Hammett equation. Happer and Wright[79] have suggested that the necessity for considering four scales of resonance parameters in applying the DSP equation (*viz.* σ_R^o, σ_R(BA), σ_R^+, and σ_R^-) could be eliminated in favour of two resonance scales and a scale of 'electron demand', relating to changes in the demand for π-electrons made on substituents. The authors propose the Variable Resonance Parameter (VRP) equation:

$$P_i - P_o = \rho_I\sigma_I + \rho_R\sigma_R^{\infty}\left[1 - \left(\frac{\sigma_R^{\infty} - \sigma_R^{H}}{\sigma_R^{\infty}}\right)^{\varepsilon}\right] \quad (3.23)$$

where ε is the electron demand parameter characteristic of the reaction, σ_R^H denotes the resonance scale for unit electron demand, and σ_R^{∞} the limiting resonance scale as $\varepsilon \to \infty$. The reader must consult the original paper for details. It is still too early to say whether the VRP equation will prove a fruitful approach to the problem of resonance scales.

Weeks and Horák[80] have proposed a new extended Hammett equation with donor and acceptor resonance contributions:

$$Q = \rho_I\sigma_I + \rho_{R+}\sigma_R^{+} + \rho_{R-}^{-}\sigma_R^{-} + h \quad (3.24)$$

It is claimed that this equation "effectively copes with wide variations of electronic effects as precisely as the formulation of Ehrenson *et al.*[29] It is also shown that the Tsuno— Yukawa (*sic*) equation is a mathematical artifact of this equation". Young and McMahon[81] have compared the treatments of certain data sets by the Yukawa— Tsuno and the Taft equations.

While Yukawa and Tsuno have in recent years continued to apply the equation with which their names are associated,[82-84] they have increasingly used a form of treatment in which inductive (σ_i) and resonance (σ_π^+ and σ_π^-) parameters are applied to the effect of *meta*- and *para*- substituents separately,[82-86] in a way analogous to Taft's DSP equation.

$$\log(k/k_o) = \rho_i\sigma_i + \rho_\pi^+\sigma_\pi^+ + \rho_\pi^-\sigma_\pi^- \quad (3.25)$$

The origins of this work are in little known papers of *ca.* 1966.[87] The work appears to have had little impact but probably merits more attention. Charton, however, considers that the parameters and equations lead to a rather incomplete separation of inductive and resonance effects.[88]

About 10 years ago Godfrey[89] proposed a multiparameter treatment which differs radically from any of the treatments discussed in the present chapter. Godfrey has developed and extended the treatment in various ways and has summarized his work to *ca.* 1977.[90] The treatment is claimed to be superior in various respects to other multiparameter treatments but has not gained the attention that it deserves, largely because many physical organic chemists find it conceptually difficult. Its very existence, however, is a salutary reminder that the analysis of substituent effects into inductive and resonance components as introduced by Taft and ordinarily practised is no more than a systematic and convenient dissection of perturbations of electron density. It is not a matter of absolute truth. Godfrey has recently applied his treatment to the properties of naphthalene and its derivatives.[91]

Krygowski and Fawcett have[92] also proposed an entirely different analysis of the substituent effect and a corresponding multiparameter extension of the Hammett equation. By using data for the entropies and enthalpies of ionization of substituted benzoic acids, the Hammett σ_m and σ_p values are separated into σ_S (for entropy) and σ_H (for enthalpy) for a number of common substituents. These are then used to analyse the log K_a values of about 30 other aromatic acid systems, through equation (3.26).

$$\log K_a = \log K_a^{\mathrm{o}} + \rho_H\sigma_H + \rho_S\sigma_S \qquad (3.26)$$

Usually entropy effects are found to make the major contribution to the total substituent effect, the entropic reaction constant being approximately equal to the normal Hammett ρ. A comparison with σ_I and σ_R values shows that entropic and inductive effects are related, while substituents with strong resonance effects are enthalpy controlled. The initial findings of this new type of correlation analysis are of considerable interest; see also a later paper.[93]

Although the application of modern sophisticated techniques of multivariate analysis is not part of the detailed subject matter of this book, we give references to the use of principal component analysis in connection with the Hammett equation and its extensions.[94-96]

Finally, an audio teaching aid for the Hammett equation has been produced.[97]

3.6 REFERENCES

1 L. P. Hammett, *J. Amer. Chem. Soc.*, *59*, 96 (1937).

2 L. P. Hammett, *Physical Organic Chemistry*, 1st ed., McGraw-Hill, New York, 1940, Chapter 7.

3 H. H. Jaffé, *Chem. Rev.*, *53*, 191 (1953).

4 O. Exner, in *Correlation Analysis in Chemistry: Recent Advances*, N. B. Chapman and J. Shorter, editors, Plenum, New York, 1978, Chapter 10.

5 C. Hansch and A. J. Leo, *Substituent Constants for Correlation Analysis in Chemistry and Biology*, Wiley, New York, 1979.

6 J. Shorter, *Correlation Analysis in Organic Chemistry*, Clarendon Press, Oxford, 1973, Chapter 2.

7 J. Hine, *Physical Organic Chemistry*, McGraw-Hill, New York, 1962, Chapter 4.

8 C. D. Johnson, *The Hammett Equation*, Cambridge University Press, 1973, Chapters 1 and 2.

9 H. C. Brown and Y. Okamoto, *J. Amer. Chem. Soc.*, *80*, 4979 (1958).

10 C. G. Swain and E. C. Lupton, *J. Amer. Chem. Soc.*, *90*, 4328 (1968).

11 H. van Bekkum, P. E. Verkade, and B. M. Wepster, *Rec. Trav. Chim.*, *78*, 815 (1959).

12 Y. Yukawa and Y. Tsuno, *Bull. Chem. Soc. Japan*, *32*, 971 (1959).

13 M. Yoshioka, K. Hamamoto, and T. Kubota, *Bull. Chem. Soc. Japan*, *35*, 1723 (1962).

14 Y. Yukawa, Y. Tsuno, and M. Sawada, *Bull. Chem. Soc. Japan*, *39*, 2274 (1966).

15 J. Shorter, in *Correlation Analysis in Chemistry: Recent Advances*, N. B. Chapman and J. Shorter, editors, Plenum, New York, 1978, Chapter 4.

16 O. Exner, in *Advances in Linear Free Energy Relationships*, N. B. Chapman and J. Shorter, editors, Plenum, London, 1972, Chapter 1.

17 M.-F. Ruasse and J.-É. Dubois, *J. Org. Chem.*, *37*, 1770 (1972).

18 R. Baker, R. W. Bott, C. Eaborn, and P. M. Greaseley, *J. Chem. Soc.*, 627 (1964).

19 L. P. Hammett, *Physical Organic Chemistry*, 2nd ed., McGraw-Hill, New York, 1970, p.365.

20 L. M. Stock and H. C. Brown, *Adv. Phys. Org. Chem.*, *1*, 35 (1963).

21 C. D. Johnson, *The Hammett Equation*, Cambridge University Press, 1973, Chapter 4.

22 C. D. Johnson, *J. Org. Chem.*, *43*, 1814 (1978).

23 J. S. Lomas and A. Argile, *J. Org. Chem.*, *44*, 2024 (1979).

24 R. Taylor, *J. Org. Chem.*, *44*, 2024 (1979).

25 R. W.Taft and I. C. Lewis, *J. Amer. Chem. Soc.*, *80*, 2436 (1958).

26 R. W. Taft and I. C. Lewis, *J. Amer. Chem. Soc.*, *81*, 5343 (1959).

27 R. W. Taft, in *Steric Effects in Organic Chemistry*, M. S. Newman, editor, Wiley, New York, 1956, Chapter 13.

28 S. Ehrenson, *Progr. Phys. Org. Chem.*, *2*, 195 (1964).

29 S. Ehrenson, R.T.C. Brownlee, and R. W. Taft, *Progr. Phys. Org. Chem.*, *10*, 1 (1973).

30 S. Ehrenson, R.T.C. Brownlee, and R. W. Taft, *Progr. Phys. Org. Chem.*, *10*, 38 (1973).

31 M. Charton, in (for example) *Correlation Analysis in Chemistry: Recent Advances*, N. B. Chapman and J. Shorter, editors, Plenum, New York, 1978, Chapter 5.

32 P. R. Wells, S. Ehrenson, and R. W. Taft, *Progr. Phys. Org. Chem.*, *6*, 147 (1968).

33 Collected data, R. Taylor, in *Comprehensive Chemical Kinetics*, Vol. 13, *Reactions of Aromatic Compounds*, C. H. Bamford and C.F.H. Tipper, editors, Elsevier, Amsterdam, 1972.

34 O. Exner, *Coll. Czech. Chem. Comm.*, *31*, 65 (1966).

35 M. Charton, personal communications. See, however, his very recent views in *Progr. Phys. Org. Chem.*, *13*, 119 (1981).

36 O. Exner and K. Kalfus, *Coll. Czech. Chem. Comm.*, *41*, 569 (1976).

37 J. D. Roberts and W. T. Moreland, *J. Amer. Chem. Soc.*, *75*, 2167 (1953).

38 C. Hansch, A. Leo, S. H. Unger, K. H. Kim, D. Nikaitani, and E. J. Lien, *J. Med. Chem.*, *16*, 1207 (1973).

39 D. F. Ewing, in *Correlation Analysis in Chemistry: Recent Advances*, N. B. Chapman and J. Shorter, editors, Plenum, New York, 1978, Chapter 8.

40 C. Hansch, in *Correlation Analysis in Chemistry: Recent Advances*, N. B. Chapman and J. Shorter, editors, Plenum, New York, 1978, Chapter 9.

41 See references in ref. 15.

42 D. H. McDaniel and H. C. Brown, *J. Org. Chem.*, *23*, 420 (1958).

43 L. D. Hansen and L. G. Hepler, *Canad. J. Chem.*, *50*, 1030 (1972).

44 S. H. Unger, Ph.D. Thesis, M.I.T., 1970.

45 N. R. Rosenquist, Ph.D. Thesis, M.I.T., 1973.

46 S. G. Williams and F. E. Norrington, *J. Amer. Chem. Soc.*, *98*, 508 (1976).

47 L. N. Patnaik, N. Mallick, M. K. Rout, and S. P. Rout, *Canad. J. Chem.*, *58*, 2754 (1980).

48 J. F. Kirsch, W. Clewell, and A. Simon, *J. Org. Chem.*, *33*, 127 (1968).

49 C. K. Hancock and J. S. Westmoreland, *J. Amer. Chem. Soc.*, *80*, 545 (1958).

50 S. I. Miller, *J. Amer. Chem. Soc.*, *81*, 101 (1959).

51 O. Exner, *Coll. Czech. Chem. Comm.*, *25*, 1044 (1960).

52 H. H. Jaffé, *J. Amer. Chem. Soc.*, *76*, 4261 (1954).

53 G. E. Dunn and T. L. Penner, *Canad. J. Chem.*, *45*, 1699 (1967).

54 G. E. Dunn and F.-L. Kung, *Canad. J. Chem.*, *44*, 1261 (1966).

55 M.J.S. Dewar and P. J. Grisdale, *J. Amer. Chem. Soc.*, *84*, 3548 (1962).

56 M.J.S. Dewar, R. Golden, and J. M. Harris, *J. Amer. Chem. Soc.*, *93*, 4187 (1971).

57 O. Exner and P. Fiedler, *Coll. Czech. Chem. Comm.*, *45*, 1251 (1980).

58 W. F. Reynolds, *J. Chem. Soc. Perkin Trans. II*, 985 (1980).

59 W. F. Reynolds, *Progr. Phys. Org. Chem.*, to appear.

60 R. Ponec, *Coll. Czech. Chem. Comm.*, *45*, 1646, 1655 (1980).

61 O. Exner, D. Ondrejičková, and J. Leška, *Org. React.*, *16*, 285 (1979).

62 D. A. Forsyth and B. B. Sandel, *J. Org. Chem.*, *45*, 2391 (1980).

63 D. F. Ewing, *Org. Magn. Reson.*, *12*, 499 (1979); this is a review of ^{13}C substituent effects in monosubstituted benzenes.

64 J. Bromilow, R.T.C. Brownlee, V. O. Lopez, and R. W. Taft, *J. Org. Chem.*, *44*, 4766 (1979).

65 J. Bromilow, R.T.C. Brownlee, D. J. Craik, M. Sadek, and R. W. Taft, *J. Org. Chem.*, *45*, 2429 (1980).

66 W. Adcock and D.P. Cox, *J. Org. Chem.*, *44*, 3004 (1979).

67 Y. Kosugi and Y. Furuya, *Tetrahedron*, *36*, 2741 (1980).

68 S. N. Sawhney and D. W. Boykin, *J. Org. Chem.*, *44*, 1136 (1979).

69 C. W. Fong, *Austral. J. Chem.*, *33*, 1291 (1980).

70 G. Butt, J. Cilmi, P. M. Hoobin, and R. D. Topsom, *Spectrochim. Acta*, *36A*, 521 (1980).

71 V. I. Glukhikh and M. G. Voronkov, *Doklady Akad. Nauk. SSSR*, *248*, 142 (1979); *English edition*, p.744.

72 T. A. Lehman and M. M. Bursey, *Ion Cyclotron Resonance Spectrometry*, Wiley, New York, 1976.

73 J. E. Bartmess, J. A. Scott, and R. T. McIver, *J. Amer. Chem. Soc.*, *101*, 6046, 6056 (1979).

74 A. Pross, L. Radom, and R. W. Taft, *J. Org. Chem.*, *45*, 818 (1980).

75 G. Kemister, A. Pross, L. Radom, and R. W. Taft, *J. Org. Chem.*, *45*, 1056 (1980).

76 J. Shorter, in *The Chemistry of the Sulphonium Group*, C.J.M. Stirling and S. Patai, editors, Wiley, Chichester, 1981, Chapter 9.

77 A. J. Hoefnagel, M. A. Hoefnagel, and B. M. Wepster, *J. Org. Chem.*, *43*, 4720 (1978).

78 I. A. Koppel, M. M. Karelson, and V. A. Palm, *Org. React.*, *11*, 101 (1974).

79 D.A.R. Happer and G. J. Wright, *J. Chem. Soc. Perkin Trans. II*, 694 (1979).

80 G. Weeks and V. Horák, *J. Org. Chem.*, *45*, 2068 (1980).

81 P. R. Young and P. E. McMahon, *J. Org. Chem.*, *45*, 5290 (1980).

82 Y. Tsuno, Y. Kusuyama, M. Sawada, T. Fujii, and Y. Yukawa, *Bull. Chem. Soc. Japan*, *48*, 3337 (1975).

83 Y. Tsuno, M. Sawada, T. Fujii, and Y. Yukawa, *Bull. Chem. Soc. Japan*, *48*, 3347 (1975).

84 Y. Tsuno, M. Sawada, T. Fujii, Y. Tairaka, and Y. Yukawa, *Bull. Chem. Soc. Japan*, *48*, 3356 (1975).

85 Y. Tsuno, M. Sawada, T. Fujii, and Y. Yukawa, *Bull. Chem. Soc. Japan*, *52*, 3033 (1979).

86 M. Sawada, M. Ichihara, Y. Yukawa, T. Nakachi, and Y. Tsuno, *Bull. Chem. Soc. Japan*, *53*, 2055 (1980).

87 Y. Yukawa and Y. Tsuno, *Mem. Inst. Sci. Ind. Res. Osaka University*, *23*, 71 (1966).

88 M. Charton, personal communication.

89 M. Godfrey, *J. Chem. Soc.*, *(B)*, 1534, 1540 (1971).

90 M. Godfrey, in *Correlation Analysis in Chemistry, Recent Advances*, N. B. Chapman and J. Shorter, editors, Plenum, New York, 1978, Chapter 3.

91 M. F. Duerden and M. Godfrey, *J. Chem. Soc. Perkin Trans. II*, 330 (1980).

92 T. M. Krygowski and W. R. Fawcett, *Canad. J. Chem.*, *53*, 3622 (1975).

93 T. M. Krygowski and W. R. Fawcett, *J. Chem. Soc. Perkin Trans. II*, 2033 (1977).

94 S. Wold and M. Sjöström, in *Correlation Analysis in Chemistry: Recent Advances*, N. B. Chapman and J. Shorter, editors, Plenum, New York, 1978, Chapter 1.

95 S. Wold and M. Sjöström, *Chem. Scripta, 2,* 49 (1972).

96 M. Sjöström and S. Wold, *Chem. Scripta, 6,* 114 (1974).

97 P. Sykes, *Linear Free Energy Relationships,* Chemistry Cassette No.11, Educational Techniques Subject Group, The Royal Society of Chemistry, London, 1981.

CHAPTER 4
The Separation of Polar, Steric, and Resonance Effects

	Page
4.1. Introduction to R. W. Taft's work	74
4.2. Applications of the Taft parameters to the reactivity of organic compounds	80
4.2.1. The linear free-energy—polar-energy relationship	80
4.2.2. The linear free-energy—steric-energy relationship	85
4.2.3. The linear free-energy—polar- and steric-energy relationship	86
4.3. Development and critique of the Taft analysis to *ca.* 1971	90
4.3.1. Further consideration of the steric parameter ..	90
4.3.2. The significance of σ*, particularly for alkyl groups	94
4.3.3. Critique of the Taft analysis	97
4.4. The Taft analysis since 1971	99
4.4.1. Steric effects	100
4.4.2. Polar effects	105
4.5. The *ortho*-effect	112
4.5.1. Taft's analysis for aromatic systems	112
4.5.2. Comparisons with *para*-substituted systems	113
4.5.3. *Ortho*-substituent constants	114
4.5.4. The work of M. Charton *ca.* 1970	115
4.5.5. Recent work on the *ortho*-effect	117
4.6. References	120

As mentioned in Chapter 1, the occurrence of steric as well as polar effects in aliphatic systems and *ortho*-substituted aromatic systems complicates correlation analysis. Little progress in developing LFER in this area was made until R. W. Taft in 1952 devised a procedure for separating polar, steric, and resonance effects, based on an analysis of the rate coefficients of basic and acidic hydrolysis of esters. Considerable use has been made of this 'Taft analysis' and of substituent parameters developed from it. The less involved applications use simple linear regression for equations analogous to the Hammett equation but the more elaborate and refined applications employ multiple regression.[1]

4.1. INTRODUCTION TO R. W. TAFT'S WORK

The kinetics of the formation and hydrolysis of carboxylic esters have been investigated extensively, and the operation of various structural effects influencing reactivity has long been recognized, *e.g.* 'steric hindrance'[2] and 'substituent polarity'.[3] The dominant influence of one particular factor is sometimes clear, *e.g.* steric hindrance is responsible for the difficulty of esterifying 2,6-disubstituted benzoic acids. In many cases, however, the simultaneous operation of several factors is conceivable and a full understanding requires their separation. A method of separating polar and steric effects in ester hydrolysis was suggested by Ingold in 1930[3] but was largely ignored until taken up by Taft *ca.* 1952.

The present account summarizes the main features of Taft's 1956 article,[4] which brought together the material of his 1952/3 papers,[5,6] with some developments. Some later work of Taft and his pupils is also included.

Taft suggested equation (4.1) for evaluating the polar effect of a substituent R in the ester RCO_2R^1,

$$\sigma^* = \left[\log(k/k^o)_B - \log(k/k^o)_A\right]/2.48 \qquad (4.1)$$

where σ^* is a polar substituent constant for R. The rate constants k refer to reactions of RCO_2R^1, and k^o to those of $MeCO_2R^1$ as standard. B and A refer to basic and acidic hydrolysis, carried out with the same R^1, solvent, and temperature. The factor 2.48 puts the σ^* values into about the same numerical range as Hammett's σ. The equation may also be applied to *ortho*-substituted benzoic esters, $o\text{-}XC_6H_4\cdot CO_2R^1$, with $o\text{-}MeC_6H_4\cdot CO_2R^1$ as standard. The terms have the following significance: $\log(k/k^o)_B$ measures the sum of the polar, steric, and resonance effects of R (or X); $\log(k/k^o)_A$ measures the sum of the steric and resonance effects of R (or X); the difference gives the polar effect of R (or X). A further equation, (4.2):

$$E_s = \log(k/k^o)_A \qquad (4.2)$$

defines a steric substituent constant, E_s, although for systems which are conjugated with CO_2R^1, E_s contains a resonance contribution.

This procedure is based on three assumptions.

(1) The relative free energy of activation, $\Delta\Delta G^{\ddagger}$, may be treated as the sum of independent contributions from polar, steric, and resonance effects.

(2) In corresponding acidic and basic reactions, the steric and resonance effects are the same.

(3) The polar effects of substituents are markedly greater in the basic than in the acidic reaction.

Assumption (1) is by no means self-evident: if the various effects interact, linear analysis is inappropriate. No progress is possible, however, unless such a simplifying assumption is made as a first approximation. Its validity is attested by the usefulness of the results obtained.

Assumption (2) lies at the heart of the analysis. The transition states (having structures closely resembling those of the carbonyl addition intermediates) for the acidic and alkaline reactions, (I) and (II), differ by two protons.

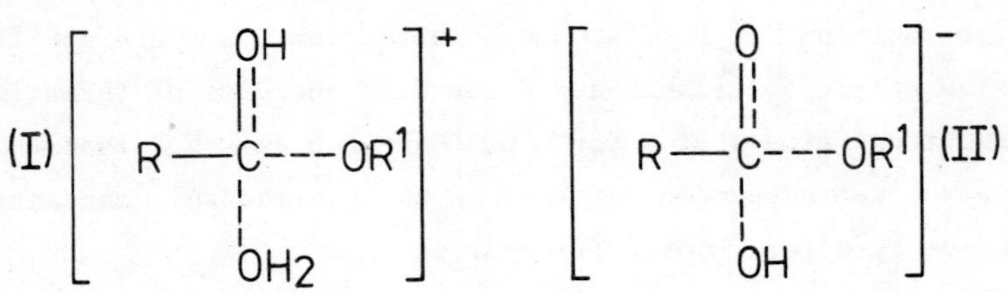

The small size of the protons makes it reasonable that "the difference in the steric interactions of the substituent, R, in the corresponding transition states for the acidic and alkaline reactions should be essentially constant, independent of R. Since both reactions involve the same initial state, the substituent steric effects in both must be very nearly equal".†[4] Further, any resonance effect between R and CO_2R^1 should be the same in the two reactions since the initial states are the same and both transition states are effectively saturated at the carboxylate carbon. Thus the steric and resonance effects of R relative to Me should be eliminated in equation (4.1). (We shall examine this assumption again later.)

Assumption (3) is supported by the Hammett ρ constants for the reactions of *meta*- or *para*-substituted benzoates.[8] For basic hydrolysis, ρ values are commonly in the range +2.2 to +2.8. For acidic hydrolysis or the related esterification of benzoic acids, recorded ρ values lie in the range 0 ± 0.5, and are often close to zero. Thus σ^* may be approximately scaled to σ through a factor of 2.48, the average of the ρ values recorded by Jaffé[8] for the saponification of benzoates, and the polar contribution to E_s may be neglected.

Taft regarded the final justification of the assumptions to be the nature of the results to which they led. He analysed a vast amount of data to give σ^* and E_s values, and used these to interpret reactivity in other reactions. For some substituents the conditions regarding constancy of solvent and R^1 could not be

† This argument is sometimes presented in a condensed form, *e.g.* "the small size of these protons makes it reasonable that the steric interaction in attaining the transition state should be the same in the two reactions".[7] It should be noted that Taft did not claim that the steric contributions to the free energies of formation of the two transition states for RCO_2R^1 are the *same* but that their *difference* is independent of R, *i.e.* he asserted the constancy of a second-order effect, not a first-order one.

completely met. Further, Taft used data from the esterification of carboxylic acids, RCO_2H, catalysed by hydrion in alcohols, R^1OH, to supplement those from acidic hydrolysis of RCO_2R^1 in aqueous organic media. This is justifiable in terms of the similarity of the transition states for the two processes; compare (I) and (III) which are inter-related through rapid proton transfers. He also made limited use of data on the alcoholysis of esters, RCO_2R^1, in the

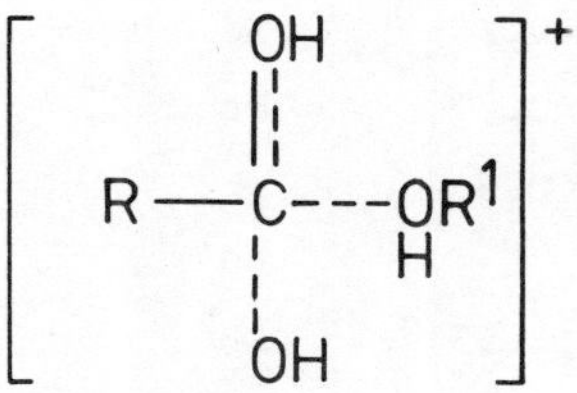
(III)

alcohol R^2OH, with the same justification. For certain substituents R, for which there was plenty of data, it appeared that $\log(k/k^o)_A$ or $\log(k/k^o)_B$ did not depend much on solvent or R^1, and that $\log(k/k^o)_A$ was approximately the same for hydrolysis and esterification. Supposedly equivalent $\log(k/k^o)$ values at 25°C, where available, were averaged, Taft holding that "the use of average values appears desirable on the basis that small specific effects and experimental errors will be reduced".[5]

Selected values of σ^* and E_s are shown in Tables 4.1 and 4.2. For a full discussion of the substituent constants for many common groups, Taft's articles may be consulted.[4-6] For more condensed discussions see Shorter.[1,7] While the E_s values in Taft's tables are always based directly on the original defining reaction, many of the σ^* values are not based directly on the hydrolysis of esters but on the application of the original σ^* values to other reactions. Table 4.1 contains various examples of this.

We give now a very brief discussion of Tables 4.1 and 4.2. Electron-withdrawing substituents have positive values of σ^* and electron-releasing substituents have negative values (Me as standard). Taft regards the values of σ^* as agreeing with "the qualitative English school theory" of the polar effect. For the electron-attracting substituents there is a recognizable relationship

TABLE 4.1. Steric and polar parameters for[a] aliphatic systems

R	σ*	$-E_s$	R	σ*	$-E_s$
Cl_3C	2.65	2.06	$Cl(CH_2)_2$	0.385[d]	0.90
Cl_2CH	1.94	1.54[h]	MeCH:CH[f]	0.36	(1.63)[g]
MeCO	1.65	-	$PhCH_2$	0.215	0.38
PhC⋮C	1.35[b]	-	PhMeCH	0.11	1.19
FCH_2	1.10[c]	0.24	$Ph(CH_2)_2$	0.08	0.38
$ClCH_2$	1.05	0.24	Me	0.00	0.00
$BrCH_2$	1.00[c]	0.27	Et	-0.10	0.07
ICH_2	0.85[c]	0.37	Pr^n	-0.115	0.36
$MeCO{\cdot}CH_2$	0.60	-	Bu^i	-0.125	0.93
Ph	0.60	(2.55)[g]	Bu^n	-0.13	0.39
$MeO{\cdot}CH_2$	0.52	0.19	Bu^tCH_2	-0.165[e]	1.74
H	0.49	-1.24	Pr^i	-0.19	0.47
PhCH:CH[f]	0.41[b]	(1.89)[g]	cyclo-C_5H_9	-0.20[c]	0.51
Ph_2CH	0.405[c]	1.76	Bu^t	-0.30	1.54

a Ref. 4.
b From catalysis of dehydration of acetaldehyde hydrate by RCO_2H.
c From ionization of RCO_2H.
d From RCO_2^- catalysis of nitramide decomposition.
e From sulphation of alcohols, ROH.
f *E* isomer.
g For significance, see text.
h Certainly highly inaccurate; see Table 4.4.(p. 101)

to the usual qualitative assessments of electronegativity of atoms and groups. For the electron-releasing alkyl groups the (negative) values of σ* show the long-accepted qualitative effects of branching at the α-carbon atom and of lengthening the carbon chain. Also the 'damping effect' of a methylene group on the influence of an electronegative substituent usually involves a decremental factor of about 2.8.

The general order of σ^*_o values (Table 4.2) corresponds very closely to that of σ_p values. When the σ^*_o values are changed to a σ_o scale with H as standard by assuming that $\sigma_o = \sigma_p$ for Me, it is seen that in various other cases $\sigma_o \approx \sigma_p$. This apparently means that the polar effects of substituents operate equally from the *ortho*- and the

TABLE 4.2. Steric and polar parameters for *ortho*-substituents in benzoates[a]

X	σ^*_o	σ_o	σ_p [b]	E_s	van der Waals radius, Å [c]
H	-	0.00	0.00	-	-
OMe	-0.22	-0.39	-0.27	0.99	-
Me	0.00	-0.17	-0.17	0.00	2.00
F	0.41	0.24	0.06	0.49	1.35
Cl	0.37	0.20	0.23	0.18	1.80
Br	0.38	0.21	0.23	0.00	1.95
I	0.38	0.21	0.27	-0.20	2.15
NO_2	0.97	0.80	0.78	-0.75	-

a Ref. 4
b D. H. McDaniel and H. C. Brown, *J. Org. Chem.*, *23*, 420 (1958).
c L. Pauling, *The Nature of the Chemical Bond*, Cornell, Ithaca, N.Y., 2nd edn., 1945.

para-position, but as a general rule this seems unlikely.

Values of σ* are important for the analysis of σ values in terms of σ_I and σ_R (Section 3.3). For a few substituents an inductive parameter σ', was based on the reactions of 4-substituted bicyclo[2.2.2]octane-1-carboxylic acids and esters.[9] The bicyclooctane system provides a good model for the study of the transmission of the non-mesomeric part of the polar effect through the benzene ring. It was found that the σ' values for X were proportional to σ* for CH_2X. This was held to confirm that σ* values were true measures of the polar effect. A new inductive parameter for substituents in general was therefore defined as $\sigma_I(X) = 0.45\sigma^*(CH_2X)$. In Chapter 3 we saw how this parameter has proved of widespread application to understanding the reactivities of *aromatic* compounds through a separation of inductive and resonance effects.

The E_s values in general conform well with qualitative assessment of 'bulk'. Increasing the bulk of the substituent leads to an increase in the steric interactions which arise when the transition state with near sp^3 hybridization at the carboxy carbon is formed from the initial state with sp^2 hybridization. Taft points out that σ* and

E_s are quite different functions of structure. This is shown very strikingly by the series Me, Et, Pr^i, Bu^t in which successive methyl groups show additivity in polar effect but an increasing increment in steric effect. Other examples of σ^* and E_s being different functions of structure may easily be recognized in Table 4.1. Note, however, that for αβ-unsaturated substituents the E_s value is often governed mainly by the conjugation of R with CO_2R^1 in the initial state; see entries in parentheses in Table 4.1. There is doubtless some resonance contribution to the E_s values for *ortho*-substituents (Table 4.2) but the values seem to reflect the expected steric effects, *e.g.* the E_s values for symmetrical *o*-X substituents parallel very well the van der Waals radii.

4.2. APPLICATIONS OF THE TAFT PARAMETERS TO THE REACTIVITY OF ORGANIC COMPOUNDS

There are three basic equations which deal respectively with the influences of polar effects alone, steric effects alone, and the two together. When necessary the equations are appropriately modified to include consideration of resonance effects. Each of these equations will be discussed in turn, although the main interest in the setting of this book will be in the equations whose application involves the use of multiple regression.

4.2.1. THE LINEAR FREE-ENERGY—POLAR-ENERGY RELATIONSHIP

Taft found that the rate or equilibrium constants for a wide variety of reactions of RY conformed respectively to equation (4.3)

$$\log (k/k^o) = \rho^*\sigma^* \qquad (4.3)$$

or the corresponding equation for equilibrium constants, where σ^* is the polar substituent constant for R and ρ^* is a reaction constant analogous to the Hammett ρ constant.† Some reactions of *ortho*-

† As with other correlation equations the writing of the combined log term on the left-hand side does not necessarily imply that the independent variable in the regression is log (reactivity relative to the standard), rather than log k, with log k^o being the statistical intercept.

substituted aromatic systems o-XC_6H_4Y also obey these equations. Examples are given in Fig. 4.1.

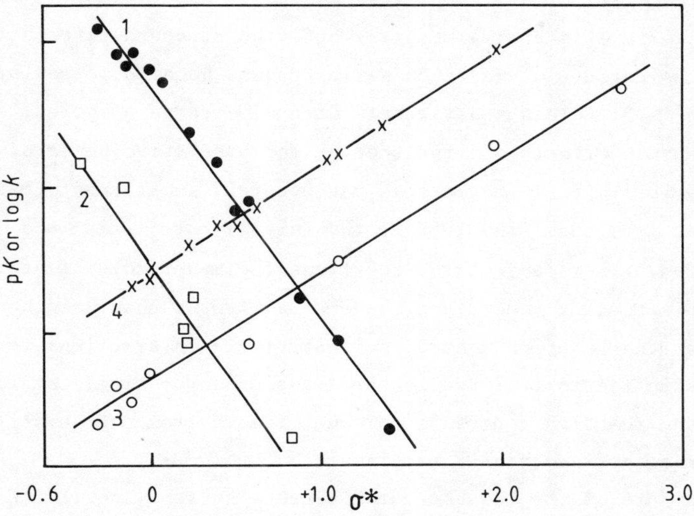

FIG. 4.1. The divisions on the ordinate are 1.00 units of *pK* or log *k* apart. The relative positions of the lines 1 to 4 with respect to the ordinate are arbitrary.
1 *pK*, aliphatic carboxylic acids, water, 25°C. (J.F.J. Dippy, *Chem. Rev.*, *25*, 151 (1939).)
2 *pK*, *ortho*-substituted benzoic acids, water, 25°C. Abscissa: Taft's σ_o values. Benzoic acid deviates markedly. (J.F.J. Dippy, *op. cit.*)
3 log *k*, alkaline hydrolysis of $[Co(NH_3)_5 \cdot O_2C \cdot R]^{2+}$, water, 25°C. (J. Basolo, J.G. Bergmann, and R.G. Pearson, *J. Phys. Chem.*, *56*, 22 (1952).)
4 log *k*, catalysis of dehydration of acetaldehyde hydrate by RCO_2H, aqueous acetone, 25°C. (R.P. Bell and W.C.E. Higginson, *Proc. Roy. Soc.*, *(A)*, *197*, 141 (1949).)

The success of such simple equations was hardly to be expected, in view of the slow progress in correlation analysis for this area compared with that for the reactivities of *meta*- and *para*-substituted benzene derivatives. However, the success of the simple equations for aliphatic and *ortho*-substituted benzene reactivities tends to be somewhat circumscribed. Conformity to equation (4.3) implies that all effects other than polar remain nearly constant throughout each reaction series. Steric effects of substituents are either completely absent or approximately the same as the steric effect of a methyl group (the standard) within the range considered. A reaction series cannot be held to conform to equation (4.3) unless substituents of a wide range of polar, steric, and resonance effects are included. The predictive power of the equation is limited: deviations may occur if substituents markedly different from those involved in the initial correlation are considered. For many of the reactions the minor role of steric effects is readily understood. Some involve no change in bond angles at the reaction centre, so that steric interactions tend to be the same in the initial and in the transition (or final) state; in others the reaction centre is somewhat remote from the substituent or the reagent involved is very small.

The failure of the unsubstituted benzene system (*i.e.* X = H) to conform to equation (4.3) in most of the aromatic reaction series indicates a constant steric or resonance effect as between the unsubstituted compound and *any* of the *ortho*-substituted compounds in a given series.

Among the great variety of processes to which equation (4.3) has been applied are the following:[10] the ionization of carboxylic acids, substituted ammonium or phosphonium ions, alcohols and thiols; the kinetic or equilibrium acidities of C—H bonds, the addition of bromine and of other reagents including free radicals to double bonds; the reactions of aldehydes and ketones; heterogeneous catalysis; polymerization; nucleophilic behaviour; and radical abstraction reactions. We shall now give some examples to illustrate how deviations from equation (4.3) were used by Taft, and subsequently by other authors, to assess other effects quantitatively, and sometimes to include the deviant systems in wider correlations through the introduction of additional parameters. Here we encounter the application of multiple regression but we defer to a later section the discussion of correlation analysis involving multiple regression on σ^* and E_s.

A good example of the assessment of steric effects is provided by the application of equation (4.3) in a form involving enthalpies instead of free energies. Equation (4.4) was applied[4] to the enthalpies of dissociation (kcal mol^{-1}) of the addition compounds formed between boron trimethyl and various aliphatic amines, $R^1R^2R^3N$.

$$\Delta H = -7.26\Sigma\sigma^* + 24.54 \qquad (4.4)$$

The term $\Sigma\sigma^*$ is the sum of the σ^* values for the groups R^1, R^2, and R^3, so that the parent system is that involving NMe_3. Ammonia and unbranched primary amines conformed well, but branched-chain compounds and secondary or tertiary amines showed marked deviations. Thus ΔH (obs.) for Bu^tNH_2 was +13.0 kcal mol^{-1}, compared to a value of ΔH (calc.) from equation (4.4) of +19.6. The difference, -6.6 kcal mol^{-1}, was attributed to steric strain in the complex. (1 kcal = 4.184 kJ.)

The study of the acidic hydrolysis of diethyl acetals of general formula $R^1R^2C(OEt)_2$ (R^1 = H or Me; R^2 variable) indicated the importance of resonance effects.

$$R^1R^2C(OEt)_2 + H_2O \xrightarrow{H^+} R^1R^2CO + 2EtOH$$

Kreevoy and Taft[11] applied equation (4.3), with $\Sigma\sigma^*$ for R^1 and R^2, to the hydrolysis of compounds derived from non-conjugated aldehydes and ketones, and obtained well-separated parallel straight lines for the two classes of compound, with a number of deviant points. A single straight line, however, was given by equation (4.5), where n

$$\log (k/k^o) = \rho^*\Sigma\sigma^* + (n - 6)h \qquad (4.5)$$

is the number of α-hydrogen atoms in R^1 and R^2, *i.e.* $(n - 6)$ is the decrease in the number of such atoms compared to the six in acetonal ($R^1 = R^2$ = Me), and h is a proportionality constant. The term $(n - 6)h$ was interpreted as a contribution from hyperconjugation in stabilizing the transition state [resembling an oxo-carbonium ion, $R^1R^2C(OEt)^+$] relative to the initial state (which is saturated).

Substituents with $\alpha\beta$-unsaturation, *e.g.* phenyl, styryl, or prop-1-enyl show large positive deviations from equation (4.5) indicating extensive resonance interaction in the transition state, *e.g.* structure (IV).

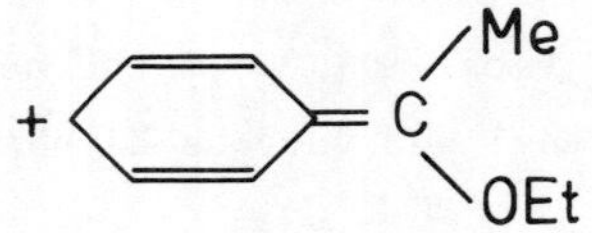

(IV)

The transition state probably has nearly the same bond angles and distances involving the central carbon atom as the initial protonated acetal molecule, $[R^1R^2C(OEt)(HOEt)]^+$. Hence steric effects should not be large. However, very large substituents may cause steric effects, *e.g.* the acetal with R^1 = Me and R^2 = neopentyl shows clear signs of steric acceleration, indicating a relief of strain when the tetrahedral initial state changes to the transition state in which the central carbon atom has made some progress towards trigonal hybridization.

Taft and Kreevoy[12] examined the application of equation (4.6) to the thermodynamics of hydrogenation of non-conjugated olefins, aldehydes, and ketones.

$$\Delta\Delta H^{\ominus} \text{ (or } \Delta\Delta G^{\ominus}) = \rho^*\Sigma\sigma^* + (n - 6)h \qquad (4.6)$$

The equation was effective for the enthalpies of gas-phase hydrogenation of 21 monosubstituted or *trans*-disubstituted ethylenes, with *trans*-but-2-ene as the standard (ρ^* = −2.41 and h = 0.44), and for the free energies of hydrogenation in toluene solution of 17 aldehydes and ketones, with acetone as the standard (ρ^* = −6.39 and h = 0.54). The greater numerical value of ρ^* for aldehydes and ketones was attributed to the carbonyl carbon atom being more electron-deficient than the carbon atoms of the olefinic double bond. By contrast the h values for the two systems agreed fairly closely, indicating that hyperconjugative stabilization is much less susceptible to electron-demand at the reaction centre. The value of h agreed fairly well with Kreevoy and Eyring's[13] theoretical model for hyperconjugation. *Cis*- or geminally disubstituted ethylenes show deviations due to steric effects, while in both systems αβ-unsaturated substituents cause additional resonance stabilization.

The work on acetal hydrolysis and on hydrogenation was re-examined by Taft and Lewis[14] as part of their work on the respective roles of C—H and C—C hyperconjugation. They concluded that the treatment of these systems would be improved by considering C—C hyperconjugation, but $h_H/h_C = 2.5$, *i.e.* C—C hyperconjugation plays only a minor part.

Equation (4.3) has also been used to detect anchimeric assistance. Streitwieser[15] showed that the solvolysis of several tertiary alkyl chlorides in 80% ethanol at 25°C conforms to equation (4.3), but $Me_2CCl{\cdot}CH_2I$ reacts about 740 times faster than expected, while for $Me_2CCl{\cdot}CH_2Bu^t$ the rate enhancement is about 15. The latter is reasonably attributed to steric acceleration by the neopentyl group, but the enormous effect of the much smaller CH_2I group indicates substantial anchimeric assistance by the iodine atom in the formation of the carbonium ion (V).

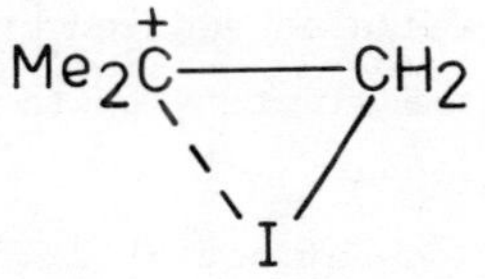

(V)

4.2.2. THE LINEAR FREE-ENERGY— STERIC-ENERGY RELATIONSHIP

Taft[4] found that a small number of reactions conformed to the equation:

$$\log (k/k^o) = \delta E_s \qquad (4.7)$$

where δ is a steric susceptibility constant. In these reactions the polar effects of substituents are evidently very small. Several of the reactions are closely related to acidic ester hydrolysis, *e.g.* the acidic hydrolysis of *ortho*-substituted benzamides, or the alcoholysis of esters, but others are very different, *e.g.* the reaction of methyl iodide with 2-alkylpyridines. In addition to correlations involving log k there are some involving enthalpy of activation or of reaction. In various examples, particularly those not resembling acidic ester hydrolysis, equation (4.7) is of rather limited applicability; deviations occur when the range of substituents is extended.

4.2.3. THE LINEAR FREE-ENERGY—POLAR- AND STERIC-ENERGY RELATIONSHIP

This is, in principle, the most general equation since the role of both polar and steric effects is considered.

$$\log (k/k^{o}) = \rho^*\sigma^* + \delta E_s \qquad (4.8)$$

It implies that the free energy of activation is a sum of independent contributions from polar and steric effects. An analogous equation may be written for equilibria. The equation constitutes a very early example of the application of multiple regression in physical organic chemistry. It was introduced by Pavelich and Taft[16] (1957) to correlate results for acid-[17] or base-catalysed[16] methanolysis of (—)-menthyl esters $RCO_2C_{10}H_{19}$ in methanol at 30°C. For the acid-catalysed reaction $\rho^*_A = 0.626$, $\delta_A = 1.549$, $n = 10$ and $R = 0.992$, while for the base-catalysed process $\rho^*_B = 2.702$, $\delta_B = 1.301$, $n = 10$, and $R = 0.996$. This work will be discussed again later. The equation was also applied by Biechler and Taft[18] to the rather complicated kinetics of the basic hydrolysis of *N*-methylanilides, PhNMe·COR.

Another early application was devised by Bowden, Chapman, and Shorter[19] (1964). The log K_a values of arylaliphatic acids with very large groups (*e.g.* Ph_2CH, Ph_3C) show no clear relationship to σ^* or E_s separately. However, multiple correlation using equation (4.8) is highly successful, and for 13 acids in 50% 2-*n*-butoxyethanol—water at 25°C yields equation (4.9).†

$$\log K_a = -6.045 + \underset{(\pm\,0.023)}{2.666}\,\sigma^* + \underset{(\pm\,0.003)}{0.252}\,E_s \qquad (4.9)$$

$$n = 13 \qquad R = 1.000 \qquad s = 0.014$$

† In 1964 the problem of collinearity of explanatory variables was not properly appreciated. The multiple regression has therefore been repeated for this book. Collinearity of σ^* and E_s is slight; $r = 0.206$. This is confirmed by the low standard errors of the regression coefficients.

Since some of the E_s values are numerically as high as *ca.* 4, the contribution of the steric term is substantial [*cf.* the successful simple regression of pK_a values for RCO_2H (water, 25°C) on σ^* for R, when only fairly small groups are involved, Fig. 4.1.] The acid-weakening steric effect was interpreted as steric inhibition of solvation of the carboxylate anion caused by the large arylaliphatic group. Later Bowden and Young[20] found a similar equation to apply to the log K_a values of 17 arylaliphatic acids in 80% 2-methoxy-ethanol—water at 25°C.

Equation (4.8) has found applications to a variety of reactions from the types previously mentioned in connection with equation (4.3), and we give now a few fairly straightforward examples.

Bowden and Hardy[21] applied equation (4.8) to the reduction of a series of 8 ketones $RCO{\cdot}CH_3$ with BH_4^- in propan-2-ol at 30°C, and found $\rho^* = 0.760$, $\delta = 0.574$, and $R = 0.998$. A considerable contribution of steric effects relative to polar effects is thus indicated. Geneste, Lamaty, and Vidal[22] studied a wider range of ketones and concluded that steric and hyperconjugative effects are important, while the inductive effect is rather small.

Equation (4.8) has recently been applied to the stability constants of the silver complexes of primary amines,[23] and to the relative rate constants for the insertion reaction of $RSiMe_2H$ with the carbene $:CHCO_2Et$.[24]

Some years ago Bogatkov, Popov, and Litvinenko applied equation (4.8) in the form of (4.10) to the nucleophilicity of amines.

$$\log (k/k^{o}) = \rho^*\Sigma\sigma^* + \delta E_N \qquad (4.10)$$

E_N is a steric constant applicable to nucleophilic reactions.[25] To arrive at E_N values they suggested that in the transition state, the amine $R^1R^2R^3N$ should be regarded as a substituent $R^1R^2R^3N^{\Delta+}$ at the electrophilic centre and taken to be isosteric with $R^1R^2R^3C$. Thus E_N values were equated with E_s values for $R^1R^2R^3C$. The applicability of equation (4.10) was examined for 18 reaction series with fair results. In a recent paper[26] the procedure is applied to the aminolysis kinetics of the vinyl chloride 1-chloro-2-*p*-nitrosulphonylethylene.

In practice, however, fairly straightforward applications of equation (4.8) are not very common. Deviations from equation (4.3) are often ascribed to steric effects but many authors do not succeed in incorporating the deviant systems in wider correlations involving steric parameters.

One factor limiting its application is the difficulty of specifying steric effects in situations where more than one substituent is involved, *e.g.* in additions to alkenes of the type (VI).

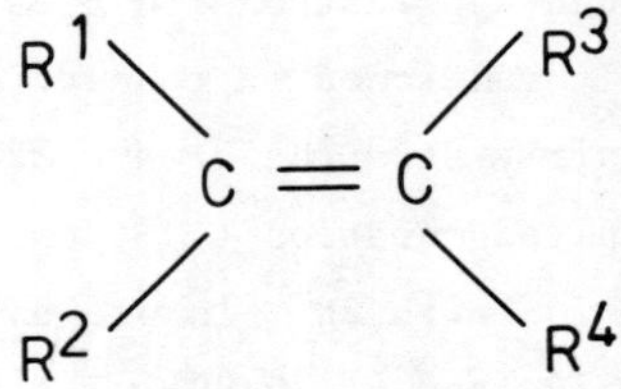

(VI)

For polar effects polysubstitution may often legitimately be treated in terms of $\Sigma\sigma^*$ (as in equations 4.4 to 4.6, and 4.10 above) but the analogous ΣE_s for steric effects is not so justifiable, *cf.* the different behaviour of E_s and σ^* in the α-branched alkyl series, Section 4.1. There have been many claims, however, that the range of correlations based on equation (4.8) is extended by including a ΣE_s term and we must examine some of these.

Dubois and his colleagues applied expressions based on equations (4.3) and (4.8) to the rates of addition of bromine to olefins in methanol (0.2 M in NaBr) at 25°C. For a wide range of olefins the influence of structure is exerted almost entirely through a polar effect. Thus for 26 olefins, whose reactivity varied by a factor of $10^{8.5}$, Dubois and Goetz[27] reported that equation (4.11) was applicable.

$$\log k_g = \log k^o + \rho^*\Sigma\sigma^* \qquad (4.11)$$

The rate coefficients, k_g, ($1\ mol^{-1}\ min^{-1}$) refer to total (*globale*) rates for Br_2 and Br_3^- as reagents, $\log k^o$ (for tetramethylethylene as parent) = 7.61, $\rho^* = -2.99$ and $r = 0.996$. The series included various (mainly hetero) substituted alkenes, $H_2C{:}CH(CHXX^1)$, in which structural variation was somewhat remote from the double bond, but also involved compounds differing considerably in the groups attached to the double bond, *e.g.* $Me_2C{:}CMe_2$ and $H_2C{:}CH_2$. Hyperconjugation does not appear to be an important factor.

For a series of 22 *hydrocarbon* olefins, *cis*-$R^1R^2C{:}CR^3R^4$, where R = H or n-alkyl, Mouvier and Dubois[28] found equation (4.11) to apply with $\log k^o = 7.75$, $\rho^* = -3.22$ and $r = 0.990$, *i.e.* $\log k^o$ and ρ^* have values much the same as for the earlier series. A further 16 compounds with branched alkyl groups deviated markedly. For the series of all 38 compounds equation (4.12) was applicable,

$$\log k_g = \log k^o + \rho^*\Sigma\sigma^* + \delta\Sigma E_s \qquad (4.12)$$

with $\log k^o = 7.42$, $\rho^* = -5.43$, $\delta = 0.96$, and $R = 0.978$, an apparently highly successful correlation. A further 32 compounds, notably *trans*- or *gem*-isomers, could not be included even in this correlation.

Despite the apparent success of equation (4.12) the bromine-addition correlations sow their own seeds of doubt. Dubois and his colleagues[27,28] have accumulated much evidence that the correct ρ^* for this reaction is *ca.* -3. When steric effects are included by using ΣE_s the value of ρ^* shifts to -5.4 but the correlation involves many olefins for which ρ^* has previously been regarded as *ca.* -3!! One could accept a small change in ρ^* but not by -2.4 units or a factor of 1.8.

Bell,[29] and Greenzaid, Luz, and Samuel[30] examined the effect of structure on K_d for the reversible hydration of aldehydes and ketones.

$$K_d = [R^1R^2CO]/[R^1R^2C(OH)_2] \qquad (4.13)$$

Bell correlated $\log K_d$ with $\Sigma\sigma^*$ and ΣE_s fairly successfully but expressed doubts about the validity of summing E_s values. Greenzaid *et al.*[30] increased the range and reliability of K_d values. Their results (12 data points) were well correlated by equation (4.14),

$$-\log K_d = \rho^*\Sigma\sigma^* + B\Delta + C \qquad (4.14)$$

where Δ is the number of 'aldehydic' hydrogen atoms, $\rho^* = 1.70$, $B = 2.03$ and $C = -2.81$, with $R = 0.995$. Correlation as proposed by

Bell[29] was much inferior, with $R = 0.928$. The Δ term implies a special stabilization of the unhydrated form on substitution of the aldehydic hydrogen atoms by alkyl groups, consistent with Dewar's theory of adjacent bond interaction.[31] Thus in this case, at least, a reinterpretation has removed the need to invoke ΣE_s.

Clotman and Zeegers-Huyskens[32] applied $\Sigma\sigma^*$ and ΣE_s to the equilibrium constants for the formation of complexes between phenol and aliphatic amines in carbon tetrachloride. Separate correlations were found for primary, secondary, and tertiary amines respectively. However, the change in ρ^* from -1.07 to -0.38 as between primary and tertiary amines seems unreasonably large.

Fischer and Funk-Kretschmar[33] correlated the rate coefficients for the Beckmann—Chapman rearrangement of ketoxime picryl ethers, $R^1R^2C{:}N{\cdot}O{\cdot}Pic$ with ΣE_s and $\Sigma\sigma^*$ for R^1 and R^2, the reactivity being controlled mainly by steric interactions in the initial state. Friedrich[34] has claimed that the pmr frequency of chloroform in the presence of sterically-hindered amines is correlated linearly with ΣE_s. Other correlations involving ΣE_s have been reported in recent years.[35-40]

Despite the empirical success of ΣE_s the physical significance of these correlations must be considered doubtful.

4.3. DEVELOPMENT AND CRITIQUE OF THE TAFT ANALYSIS TO *ca.* 1971

As previously mentioned, this area of correlation analysis was reviewed by the present author in 1970[7] and again in 1972,[1] and the reader has already been referred to these articles for more extended accounts of various topics. It is convenient at this point to summarize the accounts which were given of certain other matters in those articles, as background to a discussion of developments during the last ten years, which will be mainly deferred to Section 4.4.

4.3.1. FURTHER CONSIDERATION OF THE STERIC PARAMETER

It must be pointed out that E_s has certain features which may limit its application. First, there is the essential stereochemistry that E_s involves. E_s is based on a process in which a reaction centre with an sp^2 hybridized carbon gives rise to a transition state in which that atom is almost sp^3 hybridized. It is therefore not surprising that E_s finds its main application in systems involving the inter-relationship of sp^2 and sp^3 hybridized states, *e.g.* the hydrolysis of amides and esters, the alcoholysis of esters, reactions at the α-carbon atom of aliphatic nitro-compounds, and additions of molecules, ions, or radicals to C=C or C=O, including polymerization. Applications of E_s are not, however, restricted to systems with this characteristic, *e.g.* the use of E_s in connection with steric inhibition of solvation of carboxylate ions (as already described),[19,20] or the applications to biological activity, which are outside the scope of this book. On the other hand, E_s values seem inapplicable to the steric effects of substituents at α-carbon in the S_N2 reactions of alkyl halides with nucleophiles, which involve tetrahedrally-hybridized carbon in the initial state going to a trigonal bipyramidal transition state.[41]

Secondly, there is the question of resonance contributions to E_s. We have seen in Section 4.1 that for substituents conjugated with a carboxylic function, E_s contains a substantial resonance component. Taft recognized that even for substituents incapable of normal conjugation, there might be a contribution to E_s from a resonance interaction, *i.e.* a hyperconjugative effect of α-hydrogen atoms, *e.g.*[4]

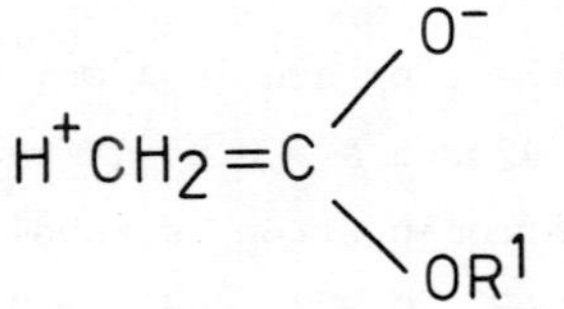

VII

In Hancock's view this should be allowed for in a 'corrected steric substituent constant', E_s^c, as in equation (4.15),[42]

$$E_s^c = E_s - h(n - 3) \qquad (4.15)$$

where h is a reaction constant for hyperconjugation and n is the number of α-hydrogen atoms. Quantum mechanical calculations by Kreevoy and Eyring[13] were used as a basis for taking $h = -0.306$ at 35°C. Selected E_s^c values are in Table 4.3. Hancock showed that the use of E_s^c values, along with other parameters in multiple regression, led to significant improvement in the correlation of the results for certain reactions.

TABLE 4.3. Steric parameters of Taft[a] (E_s), Hancock[b] (E_s^c), and Palm[c] (E_s^o)

R	$-E_s$	$-E_s^c$	$-E_s^o$
Me	0.00	0.00	0.00
Et	0.07	0.38	0.27
Pr^i	0.47	1.08	0.87
Bu^t	1.54	2.46	2.14
Pr^n	0.36	0.67	0.56
Bu^n	0.39	0.70	0.59
Bu^i	0.93	1.24	1.13

a Table 4.1.
b Equation (4.15).
c Equation (4.21).

The following relations were found to hold for the saponification of nine esters, RCO_2Me, in 40% aqueous dioxan at 35°C.[42]

$$\log k = 1.31 + 1.54\sigma^* + 0.709E_s \quad (4.16)$$
$$\log k = 1.36 + 1.48\sigma^* + 0.471E_s^c \quad (4.17)$$
$$\log k = 1.25 + 1.75\sigma^* + 0.848E_s^c - 0.383(n - 3) \quad (4.18)$$

(Rate constants in this Section are in $1\ mol^{-1}\ min^{-1}$.) For equation (4.16), $R = 0.992$ and $s = 0.076$, *i.e.* the correlation is fairly good because hyperconjugation is involved in E_s and in $\log k$. Equation (4.17) is poor ($R = 0.970$, and $s = 0.149$), because in E_s^c hyperconjugation has been eliminated. Equation (4.18) is excellent ($R = 0.998$ and $s = 0.043$) because it incorporates proper consideration of both steric and hyperconjugative effects. Statistical tests showed that the improvement in correlation of equation (4.18) over (4.16) was highly significant. The papers contain other examples.[42,43]

Hancock also showed the importance of 'change in the six-number'.[44] Newman[45] showed that the number of atoms in position 6 from the carbonyl oxygen as 1, makes a large contribution to the steric effect. When a given group is considered both as R and R^1 in RCO_2R^1 there may be a change in the six-number, Δ6, as between R and R^1. Δ6 is the six-number of a given substituent as R minus its six-number as R^1. Thus for Me, Et, Pr^i, and Bu^t, Δ6 = 0, -3, -6, and -9 respectively. Hancock used Δ6 as another structural parameter in multiple regression. For the saponification of 9 acetates, $MeCO_2R^1$, in 40% dioxan at 35°C, equation (4.19) holds,

$$\log k = 1.40 + 1.34\sigma^* + 0.730E_s^c \qquad (4.19)$$

with $R = 0.980$ and $s = 0.161$, the E_s^c values being for the substituents as R. When Δ6 is included, equation (4.20) holds,

$$\log k = 1.35 + 0.688\sigma^* + 0.664E_s^c + 0.0477\Delta 6 \qquad (4.20)$$

with $R = 0.997$ and $s = 0.070$, which is a much improved correlation. Other examples of the use of Δ6 may be found in Hancock's papers.[44,46]

Upto 1971 a few authors had used Hancock's parameters. Thus P. D. Bolton[47] regarded E_s^c and $(n - 3)$ as relevant to the acid- or base-catalysed hydrolysis of aliphatic amides and obtained significantly better correlations than with E_s. This and some later work by Bolton will be referred to in Section 4.3.3. Inaki *et al.*[48] found that the argentation constants of α- and β-alkylstyrenes could be correlated with E_s^c and $(n - 3)$ but not with E_s.

Since 1971 interest in Hancock's parameters seems to have increased, and it would be appropriate to list here a few of the applications of E_s^c *etc.*, with references: alkaline hydrolysis of thiol esters,[49] *EZ*-isomerization of 2,4-dinitrophenylhydrazones,[50] urethane formation,[51] conformational stability of alkyl vinyl ketones,[52] selectivity of alkyl radicals,[53] thermolysis of highly branched alkanes,[54] reactivity of tetrasubstituted alkenes.[55] E_s^c has also found considerable application to biological activities.[56]

Modified steric parameters were also developed by V.A. Palm,[57] who considered the contribution of both C-H and C-C hyperconjugation to E_s. Equation (4.21) was proposed:

$$E_s^o = E_s + 0.33(n_H - 3) + 0.13n_C \qquad (4.21)$$

where n_H is the number of α-C-H bonds and n_C is the number of α-C-C bonds. E_s^o is described as a 'purely steric constant', and, linked with proper consideration of hyperconjugation, has been much used by Palm and other authors in the USSR. Some values of E_s^o are in Table 4.3. Since $n_H + n_C = 3$, Palm's equation is equivalent to

$$E_s^o = E_s + 0.20(n_H - 3) \qquad (4.22)$$

and thus the difference between Hancock's and Palm's treatments essentially concerns the magnitude of the correction to be applied to E_s.

Shorter[1] summarized a few examples of the application of E_s^o. Others, both before and after 1971, are easily found in Palm's journal *Organic Reactivity*[58] or in his books.[57]

4.3.2. THE SIGNIFICANCE OF σ*, PARTICULARLY FOR ALKYL GROUPS

There seems little doubt that σ* values measure the polar effects of substituents when these are substantial. Values of σ* show good relationships to various electrical properties, *e.g.* electronegativity values, ionization potentials, dipole moments, and certain parameters of molecular orbital theory (see summary by Shorter, 1972 [1]). While such relationships tend to be dominated by the rather electronegative substituents, the σ* values for alkyl groups often fit in quite well. Well before 1971, however, the significance of small σ* values had been questioned because they might arise from an imperfect cancelling of steric effects in the Taft analysis. This applies particularly to the values for all alkyl groups, which lie mainly between 0 and -0.3. The validity of Taft's second assumption will be discussed later but it is appropriate to point out now that Taft's assertion that σ* and E_s are different functions of structure is not strictly true for alkyl groups. Thus Koppel[59] showed that for twenty primary, secondary, or tertiary alkyl groups (σ* = 0 to -0.4; E_s = 0 to -4) equation (4.23) holds, with R = 0.956 and s = 0.292, where n is the number of α-C—H bonds.

$$E_s = 0.88 + 27.78\sigma^* - 1.90(n - 3) \qquad (4.23)$$

Similarly for Palm's E_s^o, equation (4.24) holds, with $R = 0.964$ and $s = 0.265$.

$$E_s^o = 0.88 + 27.84\sigma^* - 1.71(n - 3) \qquad (4.24)$$

For electronegative substituents no such general relationships were found.

There were, moreover, other contributions casting doubt on the status of σ* values for alkyl groups. (For a more extended summary of the earlier work, see Shorter, 1972.[1]) T. L. Brown (1959)[60] discussed the possible role of polarizability effects in the influence of alkyl groups on reactivity. He concluded that such effects were so variable that the general applicability of σ* values based on ester reactions should be questioned. He went so far as to argue that Taft and Kreevoy's[12] separation of hyperconjugative effects by applying equation (4.6) might be fallacious. Ritchie and Sager (1964)[61] advanced subtle arguments to suggest that σ* values for alkyl groups are not consistent with those of other substituents and are in fact bogus, with σ* being properly zero for all alkyl groups. Shorter (1972)[1] suggested that there may be an underlying fallacy in Ritchie and Sager's arguments, but their contribution and that of T. L. Brown acquired new interest in the late 1960s because of discoveries as to the way in which alkyl groups influence gas-phase reactivity.

Modern techniques based on mass spectrometry enable gas-phase acidities and basicities to be measured. For the *acidities* of a series of amines the order

$$Bu^tNH_2 \geqslant Pr^iNH_2 > EtNH_2 > MeNH_2 > NH_3$$

was observed (Brauman and Blair 1969).[62] Interpreted conventionally in terms of the inductive effect this suggests that the *electron-attracting* power of alkyl groups *increases* with chain length and branching. On the other hand for gas-phase *basicities* orders such as

$$Me_3N > Me_2NH > MeNH_2 > NH_3$$

are observed (Munson 1965)[63] corresponding to a 'normal' electron-releasing inductive effect of methyl groups. It thus appears that in the gas-phase alkyl groups stabilize both cations and anions. In solution the stabilization of anions is not normally observed. Hence the behaviour of alkyl groups as manifested in reactions in solution must be very much influenced by interactions with the solvent. Thus it may well be naïve to suppose the σ^* values for alkyl groups are a simple measure of an inherent polar effect.

In the 1960s some indication of electron-attraction by alkyl groups in saturated systems in solution was obtained. Baker, Parish, and Stock[64] summarized the evidence for this up to 1967. It mainly concerned small effects produced by alkyl substitution remote from a reaction centre in rigid carbocyclic systems, but Schleyer and Woodforth[65] warned that steric effects could not necessarily be discounted.

The ^{13}C and ^{1}H nmr spectra of aliphatic ketones and secondary alcohols were also said to indicate an electron-attracting effect of methyl groups,[66] but this interpretation led to vigorous discussion and criticism.[67]

When the present author reviewed this area in 1972 [1] he expressed the view that "the status of σ^* values for alkyl groups may well undergo illumination and revision in the next few years. The situation is not very clear at present". Whether or not the situation is much clearer in 1981 will appear in Section 4.4.2. There has certainly been much discussion in the meantime.

Another matter which had been raised well before 1972, but which is still current, is the question of whether σ^* should be preserved as a substituent constant. Ritchie first raised this question in 1964.[61] Since $\sigma_I(X) = 0.45\ \sigma^*(CH_2X)$, correlations could equally well be carried out in terms of σ_I, leading to ρ_I, as in the application of the DSP equation to aromatic reactivity (Section 3.3). At one time Taft had himself lent his support to this view,[68] but in practice most physical organic chemists did not accept it and continued to use σ^*. In 1972 the present author pointed out[1] that while the above relationship is the basis of the σ_I scale, many of the entries in Tables of σ_I values are derived from the analysis of σ values into inductive and resonance components. It could be argued that there was some advantage in maintaining σ^* for polar substituent constants based directly on aliphatic reactivities. In 1972 Taft[68] was inclined to support this view. We shall come back to this question later.

4.3.3. CRITIQUE OF THE TAFT ANALYSIS

The Taft analysis achieved considerable empirical success and led to a deeper understanding of the influence of structure upon reactivity in a wide variety of processes, as shown in the earlier parts of this Chapter. However, Sections 4.3.1. and 4.3.2. have shown that there are questions to be answered about the significance of polar and steric parameters. By 1971 there had been specific criticisms of the basic assumptions of Taft's procedure and the present author drew attention to these.[1,7] We now summarize the main features of the earlier accounts.

Before we comment on the assumptions it is appropriate to mention one practical matter. Taft's analysis is commonly said to be a treatment of acidic and basic ester hydrolysis. In practice it was largely a treatment of acid-catalysed esterification and basic ester hydrolysis, for there was a paucity of data relating to acidic and basic ester hydrolysis in the same solvent. The reason for this is quite simple: at the high temperatures necessary for studying acidic hydrolysis, aqueous organic solvents often 'crack'. It is much easier to study acid-catalysed esterification.

Criticism of Taft's treatment has been mainly directed at assumption (2) concerning the equality of steric effects as between corresponding acidic and basic reactions. Attention has been focused in particular on the neglect of the role of the solvent: since the transition states of the acidic and basic reactions carry opposite charges, it is unlikely that the solvation patterns will be so similar that the steric interactions in the two systems will be exactly the same. Chapman, Shorter, and their colleagues were led to make this criticism as a result of various kinetic studies (see Shorter 1972 [1]), while R. M. Noyes,[69] by considering the thermodynamics of ions in solution, concluded that there are extreme differences in solvent structure around cations and anions and warned that "Taft's treatment may not have separated steric effects as clearly as was hoped".

Various other lines of evidence bear on this matter. Thus, Bowden, Chapman, and Shorter[19] examined the Taft analysis and the application of polar and steric parameters in the reactions of arylaliphatic carboxylic acids and esters (see Section 4.2.3). A Taft analysis of alkaline ester hydrolysis led to a set of σ^* values with various inconsistencies and not agreeing well with a set based on the reaction of carboxylic acids with diazodiphenylmethane. In studying conformation and reactivity, Chapman, Shorter, and Toyne[70] obtained results which also cast doubt on Taft's assumption (2). The use of a Bu^t or a $(CH_2)_4$ bridge (*trans*-decalin system) to lock the conformation of cyclohexane compounds enables the reactivity of a functional group to be studied in both the axial and the equatorial disposition. The steric interactions in the two configurations are different, and it was suggested that the procedure provides a means of changing steric effects without changing polar effects. While evidence was obtained for marked differences in steric effects as between acidic and basic ester hydrolysis in the same solvent, there were indications of a closer correspondence as between alkaline hydrolysis in aqueous organic solvents and acid-catalysed esterification in methanol.

A rather different approach to examining assumption (2) is to study the application of polar and steric parameters to reactions bearing a close resemblance to acid- and base-catalysed ester hydrolysis. The ester exchange reactions already mentioned (Section 4.2.3) in connection with equation (4.8) provide a suitable system.[16,17] For the acid-catalysed reaction $\delta_A = 1.549$, while $\delta_B = 1.301$ for the base-catalysed process, *i.e.* the steric susceptibility constants are in the ratio $\delta_A/\delta_B = 1.19$. This was attributed to the steric effects ceasing to be equivalent when they are considerably greater than those operating in ester hydrolysis ($\delta = 1$). Howsoever interpreted the results are disturbing to assumption (2). Note also that an appreciable polar effect in acid-catalysed exchange was found, *i.e.* $\rho^*_A = 0.626$.

Bolton and Jackson[47,71,72] studied the acidic and basic hydrolysis of aliphatic amides in water at 75°C. The mechanisms closely resemble those of the corresponding ester reactions. The results are best interpreted in terms of a form of equation (4.8) modified to use E_s^c values, and to include a hyperconjugative term (see Section 4.3.1). Correlation equations (4.25)[71] and (4.26)[72] finally emerged for acidic and basic reactions respectively:

$$\log k = 1.12E_s^c - 0.564(n - 3) + \log k^o \qquad (4.25)$$

$$R = 0.993 \qquad \text{(16 data sets)}$$

$$\log k = 1.67\sigma^* + 0.985E_s^c - 0.573(n - 3) + \log k^o \qquad (4.26)$$

$$R = 0.994 \qquad \text{(12 data sets)}$$

i.e. $\delta_A/\delta_B \approx 1.14$, *cf* 1.19 in ester exchange.

On the other hand support for assumption (2) came from a variant of this approach. I. V. Talvik[73] interpreted acidic and basic ester hydrolysis in *water* in terms of σ^* and E_s^o values (based on the usual measurements in organic or aqueous organic solvents), with appropriate hyperconjugative terms. She found $\delta_A \approx \delta_B \approx 0.80$.

The earlier accounts[1,7] dealt briefly with the validity of Taft's assumptions (1) and (3). In particular it was suggested that assumption (3), the basis for ignoring polar effects in the acid-catalysed reactions, required investigation. It was also pointed out that while a widespread disbelief in assumption (1) regarding the approximate separation of polar, steric, and resonance effects in a linear way would stultify study of this field, it would be appropriate to pay regard to possible contributions from interaction terms. It was said, however, that while the inclusion of cross-terms in correlations frequently improves them, their physical significance is often obscure.

4.4. THE TAFT ANALYSIS SINCE 1971

Since the author reviewed this field in the early 1970s there have been considerable developments but the main controversies continue and no definitive solutions have been found. The principal activity in the field has been in the works of Charton, DeTar, Dubois,

Grob, and Taft, and their collaborators, and in contributions by other authors who have been stimulated by the above-named. The material will be classified under the headings of steric effects and of polar effects but some of the work bears on both topics.

4.4.1. STERIC EFFECTS

Dubois has critically examined the Taft E_s scale of steric parameters. He commented unfavourably on Taft's use of a mixture of data for acid-catalysed ester hydrolysis and esterification of carboxylic acids in setting up the scale (see Section 4.1)[4-6] and suggested that it would be better to choose a *single* reference reaction.[74] In Dubois' view the best choice for this is the acid-catalysed esterification of carboxylic acids in methanol at 40°C, and on this basis he has set up a modified scale of steric parameters, which he denotes E'_s. The most reliable data for this reaction were compiled and these have been supplemented by new measurements on various acids. In all, values of E'_s for 107 substituents were presented. These include values for 44 groups originally treated by Taft; the remaining groups include many polar substituents and highly hindered alkyl groups. The E'_s scale extends over more than 8 log units. Selected values are shown in Table 4.4, along with the older E_s values where available.

In many cases E_s and E'_s do not differ greatly, although there are occasional dramatic differences where Dubois' re-examination has detected serious flaws in old data. The main interest of Dubois' work lies not so much in its role in tidying the steric parameter scale but in his subsequent discussion of the wide range of E'_s values. Considerable understanding has now been achieved of the relationship between the structure of an alkyl group and its E'_s value. The discussion is complicated and in the present account it can be dealt with only briefly, with appropriate literature citations.

In a first treatment, the steric effects of alkyl groups as characterized by E'_s have been analysed by using an approach based on the DARC† topological system and its PELCO‡ correlation method.[75] This approach involves an analysis of the systematic variation of

TABLE 4.4. Steric parameters of Taft[a] (E_s) and Dubois[b] (E'_s)

R	$-E_s$	$-E'_s$	R	$-E_s$	$-E'_s$
H	-1.24	-1.12	Ph_2CH	1.76	1.50
Me	0.00	0.00	Ph_3C	-	4.91
Et	0.07	0.08	FCH_2	0.24	0.20
Pr^i	0.47	0.48	$ClCH_2$	0.24	0.18
Bu^t	1.54	1.43	Cl_2CH	1.54	0.58
Pr^n	0.36	0.31	Cl_3C	2.06	1.75
Bu^n	0.39	0.31	$BrCH_2$	0.27	0.24
Bu^s	1.13	1.00	ICH_2	0.37	0.30
Bu^i	0.93	0.93	$PhOCH_2$	0.33	0.32
Et_2CH	1.98	2.00	Pr^iMe_2C	-	3.54
Bu^tCH_2	1.74	1.63	Et_3C	-	5.29
$PhCH_2$	0.38	0.39	Pr^i_3C	-	6.73
$Ph(CH_2)_2$	0.38	0.35	Bu^tEt_2C	-	7.21

a Ref. 4.
b Ref. 74.

E'_s in a topological sequencing of alkyl groups. The analysis leads to various successful correlations of E'_s with topology. In two further papers the treatment has been improved by the introduction of conformational considerations and much progress has been made towards understanding the contributions which the various parts of a complex alkyl group make to the E'_s value.[76,77] Some surprisingly simple relationships emerge. In the course of this work Dubois[74,77] has concluded that there is no advantage in hyperconjugative corrections of the type introduced by Hancock[42-44] and by Palm.[57,58] (See Section 4.3.1.) The importance of the six-number[44,45] as a steric parameter was, however, confirmed. Dubois has criticized a recent attempt[78] "to revive interest in the use of Hancock's E^c_s

† Description, Acquisition, Retrieval, Computer Aided Design.
‡ Propagation of an Environment which is Limited, Concentric, and Ordered.
Footnote to p. 100

parameter".[77]

For almost 10 years Fujita and his colleagues have investigated the nature and composition of the steric parameter in its E_s^c form. Thus in 1973 [79] various correlation equations were found to apply to the E_s^c values of alkyl and heteroatom-substituted groups, the most successful being:

$$E_s^c(CR^1R^2R^3) = -2.118 + 2.982E_s^c(R^1) + 2.160E_s^c(R^2) + 0.570E_s^c(R^3) \quad (4.27)$$
$$n = 37 \qquad R = 0.985 \qquad s = 0.232$$

where $E_s^c(R^1) \geqslant E_s^c(R^2) \geqslant E_s^c(R^3)$.

More recently a related approach has been used in a quantitative separation of electronic and steric substituent effects in reactions between aliphatic amines and electron acceptors,[80] *i.e.* the parametrization of the steric effect for $NR^1R^2R^3$ in terms of $E_s^c(R^1)$, $E_s^c(R^2)$, and $E_s^c(R^3)$.

Other substantial contributions to a deeper understanding of steric parameters have been made by DeTar. These have involved the calculation of steric energies by the methods of molecular mechanics. The approach has its origins in little known work by F. Becker, to which attention was drawn in one of the earlier reviews.[1] DeTar[81] first used the isoalkane $RCHMe_2$ as a structural analogue of the ester RCO_2Et and the neoalkane $RCMe_3$ as the analogue of $[RC(OH)(OH_2)OEt]^+$, the tetrahedral intermediate in acid-catalysed hydrolysis. Enthalpies of formation, ΔH_f^o, of the hydrocarbons were calculated by molecular mechanics. The differences, $\Delta\Delta H$, for the neoalkane—isoalkane pairs were found to be related to the corresponding E_s values for R by the equation:

$$E_s = -4.419 - 0.552\Delta\Delta H \quad (4.28)$$
$$n = 24 \qquad r = 0.95 \qquad s = 0.4$$

This was an encouraging start, although there were some highly deviating points. The computations were improved by calculating steric energies (by molecular mechanics) for RCO_2H as the structural analogue of RCO_2Et and for $RC(OH)_3$ as that of the intermediate.[82] The differences, ΔSE, were found to be related to the E_s values by the equation:

$$E_s = 0.340 - 0.789\Delta SE \tag{4.29}$$

$$n = 25 \qquad r = 0.98 \qquad s = 0.24$$

Thus a good account overall was given of E_s values in a range of about 4 log units, although the detailed patterns for various sub-series of groups were not well reproduced by the regression.

It must also be mentioned that DeTar has applied molecular mechanics with considerable success to steric effects in S_N2 reactions[83] and in the so-called S_N2 ring-closure reactions,[84] areas in which correlation analysis has not been particularly successful. DeTar has, however, also applied correlation analysis involving E_s to the effect of varying the state of alkylation of amine nucleophiles in S_N2 reactions (*cf.* Fujita[80] above).[85] A recent contribution by DeTar on the effects of alkyl groups on the rates of acyl-transfer reactions is very comprehensive and it will be better to deal with it mainly in Section 4.4.2.[86] Other authors have also applied molecular mechanics to the interpretation of steric effects in various reactions, *e.g.* oxidation of secondary alcohols by chromic acid[87] and reduction of ketones by borohydride.[88]

Since 1975 Charton has published a long series of papers under the general title of "Steric Effects". The most important outcome of this series has been the development of the steric parameter υ (upsilon), which has subsequently been widely discussed by Charton himself and by other authors.† Charton's work has, however, also involved a detailed examination of Taft's assumption of the equality of steric effects in acidic and basic carboxylic ester/acid reactions, leading to the conclusion that for alkyl groups in aliphatic systems the σ^* values (or σ_I values based on these) are artifacts. We shall deal with the latter aspect of Charton's work in Section 4.4.2, and will concentrate now on υ.

† The choice of the symbol υ (upsilon) has proved to be unfortunate, since it has been widely read as ν (nu) or v (vee), and then written in the latter forms by other authors.

The υ scale has its origin in Taft's comments to the effect that E_s values tend to parallel the van der Waals radii of substituents.[89] This relationship was made more quantitative by Charton in 1969.[90] Through extensive regression analysis he examined the dependence of E_s values for substituents of the types CH_2X, CHX_2, and CX_3 in aliphatic systems on van der Waals radii, and on σ_I and σ_R. Because of difficulties connected with substituent symmetry and conformation it is not possible to define a unique van der Waals radius for such groups, *cf.* monatomic substituents. This problem was handled in terms of various definitions and assumptions. A clear dependence of E_s on van der Waals radius was shown but there was no significant dependence on σ_I or σ_R. Charton regarded this as confirming that the aliphatic E_s scale really does measure steric effects.

This work was essentially the foundation for Charton's suggestion in 1973[91] that a new steric parameter υ should be defined by equation (4.30),

$$\upsilon = r_{VX} - r_{VH} = r_{VX} - 1.20 \qquad (4.30)$$

where r_{VX} is the van der Waals radius of the X group and r_{VH} is that of the hydrogen atom. This equation can be applied in a simple fashion only to monatomic substituents, *cf.* comment above. For certain substituents of the type AB_3, *e.g.* CH_3, CMe_3, $SiMe_3$, *etc.* the 'minimum' van der Waals radius $r_{V,min}$ is used.[90] For other substituents there are conformational issues. These are settled either by assumptions or by redefining υ empirically from the application of equation (4.31) to data for the acid-catalysed esterification of carboxylic acids,

$$\log k = \psi\upsilon + h \qquad (4.31)$$

the values of ψ and h having first been established by regression involving data for substituents whose υ values are based directly on equation (4.30).[92] Selected values of υ are in Table 4.5.[92] Many other values of υ and the applications of υ to various systems are in the later parts of Charton's "Steric Effects" series, easily found in *J. Amer. Chem. Soc.* and *J. Org. Chem.*, 1975 onwards.[93]

TABLE 4.5. Selected values of Charton's[a] steric parameter υ

Substituent[b]	υ	Substituent	υ
H	0.00	Et	0.56
Me	0.52	Pr^i	0.76
Bu^t	1.24	Pr^n	0.68
CF_3	0.91	Bu^n	0.68
CCl_3	1.38	Bu^s	1.02
CBr_3	1.56	Bu^i	0.98
CI_3	1.79	Bu^tCH_2	1.34
Me_3Si	1.40	Et_3C	2.38
F	0.27	$PhCH_2$	0.70
Cl	0.55	Ph_2CH	1.25
Br	0.65	CH_2Cl	0.60
I	0.78	$CHCl_2$	0.81

a Ref. 92.
b The substituents in the left-hand column have υ values based directly on van der Waals radii.

Like Dubois,[74,77] Charton[92,93] considers that the supposed hyperconjugative corrections to E_s introduced by Hancock[42-44] and by Palm[57,58] "are of very dubious validity". Dubois, however, has made some criticisms of the υ scale and considers that the implied connection of steric effect and mere bulk of substituent must break down at some point when highly hindered groups are considered.[74] DeTar[86] is critical of the way in which many υ values appear to have been derived and advocates the abandoning of υ along with Hancock's E_s^c. He is more sympathetic to Dubois' E_s' but claims that "except for values which correct erroneous experimental data, correlations with the E_s' (Dubois) set are comparable to correlations with the E_s (Taft) set".

4.4.2. POLAR EFFECTS

The question of the nature, direction, and magnitude of the polar effect of alkyl groups has continued to be much discussed in recent years.

In Part II of his "Steric Effects" series Charton[94] re-analysed the extensive data on the rate constants for basic hydrolysis of alkyl alkanoates, RCO_2R^1. He concluded that the effect of varying the alkyl group R could be adequately accounted for in terms of the steric parameter υ, with the electrical effects of the alkyl groups being unimportant. The essential difference between the effects of structure on acidic and on basic ester hydrolysis was found to lie in the values of the steric susceptibility constants [see equation (4.31)], ψ_B being considerably greater than ψ_A. It followed that Taft's assumption (2) regarding the equality of steric effects in the acidic and basic ester reactions was unwarranted and at best only an approximation. Charton concluded that the σ^* values for alkyl groups are probably artifacts and do not reflect differences in electrical effects; indeed they may actually involve steric factors. In a later paper[95] on amide hydrolyses the findings were analogous, except that there appears to be little difference between ψ_A and ψ_B.

Charton has returned to this subject again in papers on the electrical effect of alkyl groups. In the light of his findings for amide hydrolyses he examined whether it was possible to base a valid σ^* scale on such reactions.[96] The resulting values showed no relationship to those defined by Taft[4] from ester hydrolysis and no systematic dependence on alkyl group structure. These results were held to support the conclusion that Taft's σ^* constants for alkyl groups do not measure the electrical effects of such groups. In a second paper,[97] the σ^* values for alkyl groups, derived by considering various corresponding acid- and base-catalysed reactions, were successfully correlated with υ values or with n_α and n_β values, which are respectively the numbers of carbon atoms bonded to the α and β carbons of the alkyl group. In the same paper, and in another to which reference has already been made,[93] the relationship of steric effects to the branching factors n was thoroughly explored and it was concluded that an equation of the form:

$$Q = a'n_\alpha + b'n_\beta + c'n_\gamma + d'n_\delta + i \quad . \qquad (4.32)$$

is widely applicable to alkyl-group-influenced properties Q.

Distinctive regression coefficients, corresponding to different susceptibilities to the steric effects of α, β, *etc*. branching, account for the apparent need for a variety of structural effect parameters: σ^*, E_s, E_s^c, E_s^o, $(n - 6)$,† *etc*.

Taft's σ^* values for alkyl groups have been defended by Dubois[98] against some of Charton's criticisms.[92,94] Dubois' main point is that Charton's examination of σ^* for an electrical component involves the use of his own σ_I values, based on the ionization constants of acetic acids.[99] Since these σ_I values are, within experimental error, virtually identical for all alkyl groups, their use in examining σ^* seems to prejudge the issue!! Charton appears partly to have conceded this point but claims to have got round it.[96]

On the basis of his own work on the acidities of carboxamides in DMSO, Bordwell[100] has supported Charton's view that σ^* values are not valid measures of the electrical effects of alkyl groups.

DeTar[86,101] has recently re-examined many of the data for the alkaline hydrolysis of alkyl alkanoates. Like Charton he is convinced that correlation analysis on the basis of steric effects alone is satisfactory. He also endeavours "to strip off the steric component" from the σ^* values of alkyl groups to yield a "true polar residual σ^*_P". This is done on the basis of a rather poor correlation ($r = 0.81$) of σ^* with E_s for a set of alkyl groups. The σ^*_P values, however, do not bear any sensible relationship to structure. DeTar concludes that the "σ^* values of alkyl groups are an artifact; they represent a residual steric effect plus an error component".

The reality of an electronic effect of alkyl groups in aliphatic systems and, to some extent, the validity of σ^* as a measure thereof, have been vigorously defended in certain quarters. Levitt and Widing[102] have carried out an elaborate analysis of data for the gaseous ionization potentials of twenty-five series of alkyl compounds RX. A structural parameter s, which characterizes the

† *i.e.* the hyperconjugation parameter used by Kreevoy and Taft;[11] see Section 4.2.1.

effect of R on ionization potential, emerges. The s values are interpreted in terms of an electron-releasing effect of R, increasing with chain-length and -branching. When scaled to a value of -0.046 for Me, which is Taft's σ_I value for Me, the s values for other alkyl groups agree very closely with Taft's σ_I values. This appears to confirm that the σ_I values for alkyl groups, and the σ^* values on which they were based, really do measure the electronic effects of alkyl groups, and are not artifacts or measures of residual steric effects. Some values are shown in Table 4.6.[102]

TABLE 4.6. Electrical effect parameters of alkyl groups from ionization potentials[a]

R	$-s^b$	$-\sigma_I$ (Taft)	R	$-s^b$	$-\sigma_I$ (Taft)
H	0.0000	0.000	Pr^n	0.0607	0.058
Me	0.0460	0.046	Bu^n	0.0637	0.060
Et	0.0560	0.055	Bu^i	0.0657	0.064
Pr^i	0.0649	0.064	Bu^s	0.0687	0.068
Bu^t	0.0743	0.074	Am^n	0.0643	0.064

a Ref. 102.
b Scaled to 0.0460 for Me (see text).

Taft and Levitt[103] have subsequently used the ionization-potential-based σ_I values in the correlation of the gas-phase basicities of primary aliphatic amines. In this paper attention is paid to the role of both an inductive (polarization) effect and a polarizability effect in governing the electronic behaviour of alkyl groups. (*cf*. T. L. Brown, 1959;[60] see Section 4.3.2). The problem is that these are likely to show very similar structural dependences. Also, in some processes inductive and polarizability effects will reinforce, while in others they will be opposed. Polarizability effects may be expected to dominate inductive effects for gas-phase proton-transfer processes. All this is exemplified in a second paper by Taft *et al.*,[104] which purports to make a separation of polarizability and inductive effects in certain gas- and solution-phase proton-transfer equilibria. Thus the reverse order of acidities of aliphatic alcohols in the gas-phase and in hydroxylic solvents is explained, since polarizability effects determine the gas-phase acidity order but inductive effects of the opposite order (reduced but not eliminated by solvation) determine the solution acidity order. The σ_I values of alkyl groups (and certain polar substituents) are corrected for a small polarizability component. Structural effects on gas-phase and solution acidities are also discussed in a later paper.[105]

Screttas has examined the use of the ionization potentials of free radicals as alkyl inductive substituent constants,[106] and the correlation of activation energies with Taft's alkyl inductive substituent constants.[107] The latter paper leads "to the conclusion that a given set of kinetic data can be represented by a number of self-consistent but nevertheless arbitrary ways ..... it is felt that Charton's questioning of the validity of the σ^* and $\sigma_I(R)$ scales is unjustified".[108]

The apparent electron-attracting effect of alkyl groups sometimes revealed by nmr studies[66,67] has been discussed again under the title of "Non-equivalence of the various criteria for alkyl inductive effects".[109] The measurement of intrinsic inductive substituent effects by nmr has also been discussed by Adcock and Khor.[110]

In Section 4.3.2. it was mentioned that the need to retain σ^* as a substituent constant has long been questioned. Nevertheless σ^* still survives on active service in physical organic chemistry,[111] although those who advocate abandoning it in favour of σ_I are vociferous.[112]

It is an undoubted fact that the σ_I scale was based on σ^* through the equation $\sigma_I(X) = 0.45\sigma^*(CH_2X)$ (Section 4.1), the scaling factor being considered appropriate for the use of σ_I in the separation of inductive and resonance effects in Hammett σ_m and σ_p values (Section 3.3). Also it is a fact that the σ_I scale began at an early stage to win some degree of independence from σ^*. On the one hand there was a tendency to adjust σ_I values to give the best analyses of Hammett σ values. On the other hand, many σ_I values were not based on the original Taft analysis but on the application of the first σ*-based σ_I values (primary values) to other processes, notably the ionization of aliphatic or alicyclic carboxylic acids. Charton's 1964 paper reveals this very clearly.[99] In his latest and most extensive article on "Electrical effect substituent constants for correlation analysis" the σ* basis of σ_I is abandoned altogether.[112] Primary σ_I values are now *defined* by Charton from the ionization constants of 4-substituted bicyclo[2.2.2]octane-1-carboxylic acids in 50% w/w EtOH-H_2O at 25°C by the equation $\sigma_I = \Delta pK_a/1.56$.

In view of the doubts about the effectiveness of Taft's separation of polar and steric effects in ester hydrolysis, this separation of σ_I from its original σ* base is probably wise. This development makes relatively little difference to the σ_I values of fairly polar substituents, but for alkyl groups it produces values which lie mainly in the region 0.00 ± 0.01 and bear no discernible relationship to structure. It should be mentioned, however, that a rival set of inductive constants has been based by C. A. Grob and his colleagues on the ionization of 4-substituted quinuclidinium ions (VIII) in water.

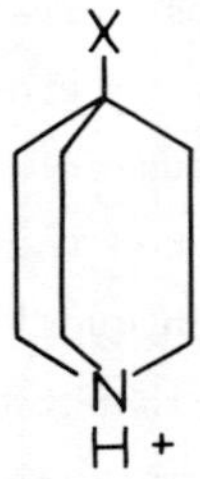

VIII

These are designated σ_I^q, defined simply as $pK_a^H - pK_a^X$. Because the system is highly sensitive to substituent effects, the numerical scale of σ_I^q is rather extended. Selected values are in Table 4.7, with σ_I values from Charton[112] for comparison. The quinuclidine system is able to distinguish between the electrical effects of alkyl groups in the α-branched series, finding the order $Bu^t > Pr^i > H > Et > Me$ for electron-releasing ability. Thus the alkyl groups are in the order of electron release given by σ* but H is out of place. It seems possible that this is due to some constraint in the ring system on replacing H by a more bulky substituent. Applications of σ_I^q to the reactions of a variety of systems have been examined by Grob in a series on "Polar Effects".[114]

TABLE 4.7. Inductive constants based on the quinuclidine system[a]

X	σ_I^q	σ_I(Charton)[b]	X	σ_I^q	σ_I(Charton)[b]
H	0.00	0.00	COMe	1.69	0.30
Me	0.11	-0.01	CN	3.04	0.57
Et	0.03	-0.01	NO_2	3.48	0.67
Pr^i	-0.08	0.01	OH	1.68	0.24
Bu^t	-0.15	-0.01	OMe	1.81	0.30
CH_2OMe	0.66	0.11	F	2.57	0.54
CH_2Cl	0.97	0.17	Cl	2.51	0.47
Ph	0.94	0.12	Br	2.65	0.47
CO_2Me	1.70	0.32	I	2.34	0.40

a Ref. 113.
b Ref. 112.

Do the Taft analysis equation (4.1) and the validity (or otherwise) of his assumption regarding the equality of steric effects in acidic and basic ester reactions matter much any longer? No simple answer can be given, and the present author and his colleagues felt that Taft's assumption was worth some further study by means of conformationally-locked cyclohexane systems, *cf.* the earlier work.[70] This further study[115] involved the methoxy-exchange reactions of methyl *trans*-decalin-2α- and 2β-carboxylates in CD_3OD, under acidic or basic conditions. The ratios of rate constants for the reactions of the *eq* and *ax* compounds varied with reaction conditions but under comparable conditions were similar for the acidic and basic reactions. It was concluded that in these systems Taft's assumption holds with reasonable precision.

Finally it should be mentioned that the exponents of elaborate statistical techniques have turned their attention to aliphatic substituent effects. Sjöström and Wold[116] have applied principal component analysis to alkyl group effects and have concluded that a two-term relationship is needed, *i.e.* that alkyl groups exert some effect in addition to a steric effect. The choice of data for this analysis has, however, been strongly criticized by DeTar.[86]

4.5. THE *ORTHO* EFFECT

Steric phenomena have long been recognized as playing a major part in the peculiar effects of *ortho*-substituents. Primary steric effects of various kinds, including steric hindrance to solvation or to the approach of the reagent, and secondary steric effects have been invoked. In certain systems hydrogen-bonding and other intramolecular interactions have been postulated.[117,118] The main approaches to understanding the *ortho*-effect have involved attempts to specify quantitatively the polar effect and then to draw inferences about the operation of other factors. Only in recent years has a more complete analysis through multiple regression on appropriate parameters become possible in some cases.

4.5.1. TAFT'S ANALYSIS FOR AROMATIC SYSTEMS

Values of polar and steric parameters for *ortho*-substituents have been discussed in Sections 4.1 and 4.2. Taft himself applied σ^*_o values or σ_o values (H as standard) to several reaction series by using equation (4.3). The parent system (X = H) tended to deviate, suggesting a difference in steric or resonance effect as between the parent compound and any of the substituted compounds. Further, ρ^* sometimes differed markedly from ρ for the corresponding *meta*- or *para*-substituted systems (*e.g.* $\rho^* = 1.79$, $\rho = 1.00$ for ionization of benzoic acids in water), although $\rho^* \approx \rho$ by definition for the alkaline hydrolysis of benzoates.

Compared with the aliphatic parameters, the Taft parameters for *ortho*-substituents have subsequently found relatively few applications. In various studies the use of Taft's σ_o values serves to place *ortho*-substituted compounds on the Hammett plot for *meta*- and *para*-derivatives, *e.g.* Little *et al.* simultaneously correlated the chronopotentiometric quarter-wave potentials, $\Delta E_{\frac{1}{4}}$, for 8 *ortho*- and 41 *meta*- or *para*-substituted phenylferrocenes by applying Taft's σ_o values and σ_m or σ_p values respectively.[119] Taft's E_s values for *ortho*-substituents have been used in qualitative discussion but precise use of them is rare. An exceptional application was of a form of equation (4.8) to the rate of reaction of benzenediazonium ions with six benzaldehyde *ortho*-substituted phenylhydrazones to give formazans.[120] Values of ρ_o and δ were -1.3 and 1.06 respectively, indicating substantial contributions from both polar and steric effects. (See Sections 4.5.4. and 4.5.5. for further comments on Taft's parameters for *ortho*-substituents.)

4.5.2. COMPARISONS WITH *PARA*-SUBSTITUTED SYSTEMS

The idea that *para*-substituted systems provide a basis for interpreting the *ortho*-effect goes back to 1928, when Kindler attempted to evaluate steric effects in the alkaline hydrolysis of *ortho*-substituted ethyl benzoates by comparing the behaviour of corresponding benzoates and cinnamates.[121] The approximate equality of σ_p and Taft's σ_o values for several substituents has encouraged various authors to use an appropriate *para*-substituent constant (σ_p, σ_p^o, σ_p^+, *etc.*) as a measure of the polar effect of an *ortho*-substituent. The additional structural influences constituting the *ortho*-effect may then be revealed. Similarly values of (k_o/k_p) or $\log (k_o/k_p)$ have been taken as a basis for assessing an *ortho*-effect.

Shorter (1972)[1] compiled various examples of these approaches, from which we now repeat just two.

Eaborn *et al.* studied the effect of *ortho*-substituents in the protodesilylation of $XC_6H_4{\cdot}SiMe_3$.[122] A Yukawa-Tsuno function $[\sigma_p + 0.3(\sigma_p^+ - \sigma_p)]$ (see Section 3.2) correlated the results for most of 14 compounds. The coefficient, 0.3, is smaller than that needed for *para*-substituents (0.7), indicating that the $-R$ effect is less important relative to the inductive effect for *ortho*-substituents. Steric effects were evidently small.

Watkinson *et al.* examined the variation of $\log (k_o/k_p)$ with solvent composition for the saponification of ethyl phenylacetates in ethanol-water mixtures, and found evidence for steric inhibition of solvation of the transition state in the reactions of the *ortho*-substituted esters.[123]

4.5.3. *ORTHO*-SUBSTITUENT CONSTANTS

Many authors have selected a reaction believed free from the steric effects of *ortho*-substituents and have derived a scale of σ_o values by assuming that $\rho_o = \rho$ for this reaction. For instance McDaniel and Brown[124] believed that pK_a values for 2-substituted pyridinium ions provided a basis for σ_o values. Usually, however, systems in which the reaction centre is somewhat removed from the ring have been chosen, *e.g.* the dissociation of arylpropiolic acids, ArC≡C·CO_2H.[125] Values of σ_o have also been based on spectroscopic data.[126] In 1969 Charton assembled 32 sets of *ortho*-substituent constants.[127] A selection is in Table 4.8.

TABLE 4.8. Polar constants for *ortho*-substituents

Probable status	OMe	Me	F	Cl	Br	I	NO_2
σ *a*	-0.39	-0.17	0.24	0.20	0.21	0.21	0.80
σ *b*	-0.13	-	0.13	0.24	-	-	0.55
σ *c*	-	-0.14	0.53	0.83	0.91	0.84	-
σ^o *d*	-0.53	-0.16	0.16	0.31	-	-	0.94
σ^o *e*	-0.67	-0.14	0.23	0.37	0.41	0.43	0.97
σ^+ *f*	-0.43	-0.25	-	0.45	0.55	-	0.75
σ^- *g*	0.00	-0.13	0.54	0.68	0.70	0.63	1.24
σ^- *h*	-0.37	-0.13	0.29	0.50	0.55	0.64	1.20

a From Taft's σ_o^* values. Table 4.2. Ref. 4.
b Ionization of phenylpropiolic acids in 35% dioxan. Ref. 125.
c Ionization of benzoic acids in water, other effects having been eliminated. Ref. 132.
d Pyrolysis of isopropyl benzoates. D.A.K. Jones and G.G. Smith, *J. Org. Chem.*, *29*, 3531 (1964).
e Association of benzoic acids with 1,3-diphenylguanidine. M.M. Davis and H.B. Hetzer, *J. Res. Nat. Bur. Stand.*, *60*, 569 (1958). Calculated by D.A.K. Jones and G.G. Smith, *op. cit.*
f Mean values from pyrolysis of arylethyl benzoates, acetates, and methyl carbonates. G.G. Smith, K.K. Lum, J.A. Kirby, and J. Posposil, *J. Org. Chem.*, *34*, 2090 (1969).
g Ionization of phenols. G.B. Barlin and D.D. Perrin, *Quart. Rev.*, *20*, 75 (1966).
h Pmr shifts of phenols in DMSO; *o*-NO_2 compound in HMPT. Ref. 126.

There is often poor agreement between the various values determined for a given substituent. It is, however, important to remember that the defining processes are very varied in nature and in consequence σ_o values may variously be the equivalent of σ, σ^o, σ^+, σ^-, *etc.* In some cases the defining process is probably subject to unsuspected steric or other interference from the *ortho*-substituent. If one makes allowance for all these causes of discrepancies, the agreement still seems poor and there appear to be real difficulties in the way of determining satisfactory σ_o values. This situation has prompted a search for the 'true' *ortho*-substituent constant, but the complexity of the influence of *ortho*-substituents on reactivity may make the search for σ_o scales of wide applicability quite fruitless. Charton has marshalled much evidence to support such a gloomy view.[127,128]

4.5.4. THE WORK OF M. CHARTON *ca.* 1970

Only the main findings can be given here. Charton (1971) has written a lengthy summary article.[128] The work was based on the separation of the polar effect of a substituent into inductive and resonance contributions through a dual-parameter treatment involving correlation equations of the general form

$$Q = \alpha\sigma_I + \beta\sigma_R + h \qquad (4.33)$$

which Charton refers to as the extended Hammett equation. Q is the property being correlated (log k, log K, σ_o, *etc.*), h is the intercept term, and α and β define the relative importance of inductive and resonance effects. In each application Charton examined the significance of introducing an additional term to represent the steric effect of the substituent as a function of van der Waals radius. The extra term was written as ψr_V.†

Charton's main findings were as follows:

† The symbols used by Charton have varied. In some work λ and δ have been used instead of α and β. Also Charton often adds X as subscript to indicate that the effect of substituent X is under consideration, *e.g.* $\sigma_{I,X}$, $\sigma_{R,X}$, $r_{V,X}$. The present author regards this as an unnecessary complication for the discussion in this Chapter. Later, r_V was the basis of υ; see Section 4.4.1.[91]

(1) Most of the proposed sets of σ_o values could be adequately analysed in terms of σ_I and σ_R-type parameters, but the ratio β/α varied over the range 0.3 to 1.3. Further, h varied from 0.78 to -0.27 indicating that the standard $\sigma_o = 0.0$ for H is not satisfactory. Charton concluded that no completely general set of σ_o values could be defined; nor did it appear likely that a small number of scales, each for a particular type of situation, would prove successful.

(2) A vast amount of data relating to the *ortho*-effect was analysed successfully by the above equation without including the steric effect term. Charton concluded that steric effects play a minor role in the *ortho*-effect and that different contributions of resonance and inductive effects, *i.e.* β/α values, account mainly for the variety of phenomena. This even applies to the E_s values of *ortho*-substituents, which in Charton's view are thus not steric parameters at all!! (For substituents in aliphatic systems, however, Charton found that E_s was essentially a function of substituent size, with σ_I and σ_R terms making insignificant contributions, *i.e.* E_s is indeed a steric parameter. See Section 4.4.1.[90]) Rates of alkaline hydrolysis and of methanolysis of 2-substituted benzoates are largely a function of σ_I. On the other hand, the structural effect in acid-catalysed esterification of benzoic acids or the acid-catalysed hydrolysis of benzoates is largely a function of σ_R. For the ionization of *ortho*-substituted benzoic acids in various media σ_I and σ_R make comparable contributions.

In summarizing Charton's work Shorter (1972)[1] pointed out that Charton's data sets frequently lacked substituents whose steric effect is expected to be substantial. Often the necessary items were not available but sometimes bulky substituents, *e.g.* Bu^t or I were admitted not to conform. It was suggested that the exclusion of data relating to bulky groups might distort the picture for the less bulky substituents appearing in the correlation. Shorter also described as an "awkward matter" the frequent need to exclude the parent compound as showing a marked deviation. He commented "The correlation of data made possible by computers is valuable but one must not lose sight of the chemistry in a welter of statistics. Real effects can be obscured by good correlation coefficients and high confidence levels. The success of a correlation expression in reproducing the detailed pattern of the data is also relevant."

4.5.5. RECENT WORK ON THE *ORTHO*-EFFECT

Charton's work on the *ortho*-effect has continued in association with the present author and his colleagues, who have measured rate coefficients for the reactions of benzoic acid and 32 *ortho*-substituted benzoic acids with diazodiphenylmethane (DDM) at 30°C in 11 alcohols, including 2-methoxyethanol.[129] For the reactions in a given alcohol the correlation equation (4.34)

$$\log k = \alpha\sigma_I + \beta\sigma_R + \phi\upsilon + h \qquad (4.34)$$

was applied with a limited set of substituents (up to 18), whose σ_I, σ_R, and υ parameters are well established (see Section 4.4 [92,112]) and are unlikely to show effects due to hydrogen-bonding or substituent conformation. The results for other substituents were then interpreted by comparing log k(calc.) from a correlation expression with log k(obs). A very detailed and illuminating analysis was possible. A more limited study involving the reactions of *ortho*-substituted benzoic acids and DDM in aprotic solvents was also carried out.[130] Charton's steric parameter υ gave a good account overall of the secondary steric effects of the *ortho*-substituents. On the basis of older work of Chapman, Shorter, and colleagues[131-2] Charton had previously conceded that this system shows "the most certain, clear-cut steric effect we have ever observed for any physical or chemical data for *ortho*-substituted benzenes" (*cf.* Section 4.5.4).[133] One of the most striking findings was the very great importance in aprotic solvents of effects involving hydrogen-bonding between *ortho*-substituent and reaction centre.

Some of Charton's conclusions in his work on the *ortho*-effect *ca.* 1970 (Section 4.5.4) have very recently been criticized by authors in the GDR. Thus H. Mager *et al.*[134] claim to have shown that Taft's E_s values for *ortho*-substitutents *are* functions of steric effects, and not merely of inductive and resonance effects as asserted by Charton.[90,128] The main point seems to be that interpretation of Taft E_s values for *ortho*-substituents depends strongly on the nature of the substituents included in the analysis. The matter has been pursued very thoroughly by H. Mager[134] and by P. P. Mager,[136] by using a variety of statistical techniques. H. Mager[137] also claims that steric effects are approximately the same in acidic and basic *ortho*-substituted ester reactions, *i.e.* that Taft's assumption (2) is valid for such systems.

Godfrey has applied his FCT treatment of substituent effects (Section 3.5) to the influence of *ortho*-substituents on side-chain reactivity.[138] Krygowski and Fawcett have found that their extended Hammett equation, based on a separation of enthalpic and entropic contributions to the total substituent effect (Section 3.5), may be used to correlate certain *ortho*-substituted systems.[139]

Fujita and Nishioka[140] have devoted much effort to devising a treatment of the *ortho*-effect based on the Swain—Lupton treatment (see Section 3.4). The authors seek to include *ortho*-substituted systems in the same correlation equations as *meta*- and *para*-substituted systems. The total *ortho*-substituent effect is assumed to be composed of an "ordinary polar effect" and "proximity effects". The former is defined as equal to that of *para*-substituents and thus $\sigma_o \equiv \sigma_p$, $\sigma_o^+ \equiv \sigma_p^+$, *etc.* The proximity polar effect is considered proportional to the $\mathfrak{F}$ parameter, while the steric effect is given the Taft E_s constant for the substituent when in an aliphatic system. The treatment cannot parametrize hydrogen-bonding and other specific intramolecular interactions. The basic equation is (4.35),

$$\log k_{o,m,p} = \rho\sigma_{o,m,p} + \delta E_s^o + f\mathfrak{F}_o + C \qquad (4.35)$$

with the E_s and $\mathfrak{F}$ terms being necessarily zero for *meta*- and *para*-substituents. $\sigma_{o,m,p}$ implies a choice of sigma scale appropriate to the system.

The authors applied their equation to numerous data sets from the literature and to their own results on the alkaline hydrolysis of substituted phenyl esters.[141] The correlation equations often appear to be statistically very satisfactory. The implied equality of resonance effects for *ortho*- and *para*-substituents may be adversely criticized, particularly as special σ_o values have sometimes to be used, evidently to deal with the resonance effects *not* being equal in certain cases. It should also be pointed out (as the authors do) that correlation with a σ scale and $\mathcal{F}$ is equivalent to correlation with $\mathcal{F}$ and $\mathcal{R}$. Hence the treatment could be reformulated as correlation on $\mathcal{F}$, $\mathcal{R}$, and a steric term. It is thus subject to the same criticisms as the Swain—Lupton treatment of *meta*- and *para*-substituent effects (Section 3.4.2). The treatment is also, of course, closely related to Charton's treatment involving equation (4.34). However, the attempt to include *ortho*-, *meta*- and *para*-systems in one equation is an interesting and novel (albeit possibly restricting) feature.

Consiglio *et al.*[142] have recently applied the Fujita—Nishioka treatment to the rate constants for anilino-debromination of 2-bromo-3,5-dinitrothiophen (IX) by aniline, and six *ortho*-, five *meta*-, and eight *para*-substituted anilines.

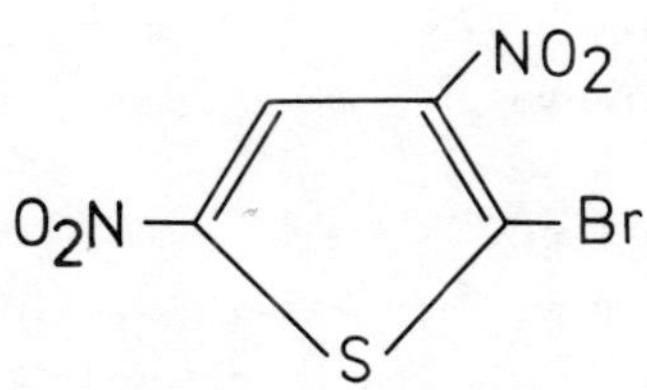

IX

The regression coefficients obtained were: $\rho = -2.96$, $\delta = 1.31$, $f = -1.83$, with $s = 0.09$ and $R = 0.998$. The same research group has studied the ionization of 3-substituted thiophen-2-carboxylic acids[143] and the alkaline hydrolysis of the corresponding methyl esters.[144] The Yukawa—Tsuno equation (Section 3.2) was applicable to both sets of results, without any need for a steric term.

Detailed discussions of the *ortho*-effect have also been presented for the ionization of several series of *cis*-β-substituted α,β-unsaturated carboxylic acids (with substituents H, Me, Et, Pr^i, Bu^t)[145] and of 2-substituted-1-azuloic acids.[146]

Berg *et al.* have recently based a new scale of *ortho*-steric parameters S^{o} on the kinetics of quaternization by methyl iodide of a large number of *ortho*-substituted pyridines.[147] ^{13}C nmr studies of the *ortho*-effect have also been carried out.[148,149]

4.6 REFERENCES

1 For detailed discussion of the subject matter of this Chapter to *ca.* 1971, see J. Shorter, in *Advances in Linear Free Energy Relationships*, N. B. Chapman and J. Shorter, editors, Plenum, London, 1972, Chapter 2.

2 V. Meyer and J. J. Sudborough, *Chem. Ber.*, *27*, 1580 (1894).

3 C. K. Ingold, *J. Chem. Soc.*, 1032 (1930).

4 R. W. Taft, in *Steric Effects in Organic Chemistry*, M. S. Newman, editor, Wiley, New York, 1956, Chapter 13.

5 R. W. Taft, *J. Amer. Chem. Soc.*, *74*, 2729, 3120 (1952).

6 R. W. Taft, *J. Amer. Chem. Soc.*, *75*, 4231, 4534, 4538 (1953).

7 J. Shorter, *Quart. Rev.*, *24*, 433 (1970).

8 H. H. Jaffé, *Chem. Rev.*, *53*, 191 (1953).

9 J. D. Roberts and W. T. Moreland, *J. Amer. Chem. Soc.*, *75*, 2167 (1953).

10 Detailed examples and references in refs. 1 and 7.

11 M. M. Kreevoy and R. W. Taft, *J. Amer. Chem. Soc.*, *77*, 5590 (1955).

12 R. W. Taft and M. M. Kreevoy, *J. Amer. Chem. Soc.*, *79*, 4011, 4016, (1957).

13 M. M. Kreevoy and H. Eyring, *J. Amer. Chem. Soc.*, *79*, 5121 (1957).

14 R. W. Taft and I. C. Lewis, *Tetrahedron*, *5*, 210 (1959).

15 A. Streitwieser, *J. Amer. Chem. Soc.*, *78*, 4935 (1956).

16 W. A. Pavelich and R. W. Taft, *J. Amer. Chem. Soc.*, *79*, 4935 (1957).

17 W. A. Pavelich, Thesis, Pennsylvania State University, 1955.

18 S. S. Biechler and R. W. Taft, *J. Amer. Chem. Soc.*, *79*, 4927 (1957).

19 K. Bowden, N. B. Chapman, and J. Shorter, *J. Chem. Soc.*, 5239 (1963); 3370 (1964).

20 K. Bowden and R. C. Young, *Canad. J. Chem.*, *47*, 2775 (1969).

21 K. Bowden and M. Hardy, *Tetrahedron*, *22*, 1169 (1966).

22 P. Geneste, G. Lamaty, and B. Vidal, *Compt. Rend.*, *(C)*, *266*, 1387 (1968).

23 R. D. Hancock, *J. Chem. Soc. Dalton Trans.*, 416 (1980).

24 L. P. Danilkina, O. V. Dorokhova, A. F. Marchenko, and Ya. M. Slobodin, *Zh. Obshch. Khim.*, *49*, 1259 (1979); *English edition*, p.1105.

25 S. V. Bogatkov, A. F. Popov, and L. M. Litvinenko, *Reakts. Sposobnost. Org. Soedin.*, *6*, 1011 (1969); *English edition*, p.436.

26 A. Popov, V. Kravchenko, Zh. Piskunova, and L. Kostenko, *Org. React.*, *16*, 325 (1979).

27 J.-E. Dubois and E. Goetz, *J. Chim. Phys.*, *63*, 780 (1966).

28 G. Mouvier and J.-E. Dubois, *Bull. Soc. Chim. France*, 1441 (1968).

29 R. P. Bell, *Adv. Phys. Org. Chem.*, *4*, 1 (1966).

30 P. Greenzaid, Z. Luz, and D. Samuel, *J. Amer. Chem. Soc.*, *89*, 749 (1967).

31 M.J.S. Dewar and H. N. Schmeising, *Tetrahedron*, *5*, 166 (1959).

32 D. Clotman and Th. Zeegers-Huyskens, *Spectrochim. Acta*, *23A*, 1627 (1967).

33 H. P. Fischer and F. Funk-Kretschmar, *Helv. Chim. Acta*, *52*, 913 (1969).

34 H. J. Friedrich, *Angew. Chem. Int. Edn.*, *4*, 697 (1965).

35 Y. Ogata and S. Watanabe, *J. Org. Chem.*, *45*, 2831 (1980).

36 L. Červený, J. Bartoň, and V. Růžička, *Sci. Papers Prague Inst. Chem. Tech.*, *C24*, 125 (1976).

37 H. G. Hauthal, H. Onderka, and W. Pritzkow, *J. Prakt. Chem.*, *311*, 82 (1969).

38 H. J. Bergmann, G. Collin, G. Just, G. Müller-Hagen, and W. Pritzkow, *J. Prakt. Chem.*, *314*, 285 (1972).

39 W. M. Baluzow, G. Just, and W. Pritzkow, *J. Prakt. Chem.*, *316*, 1051 (1974).

40 W. M. Baluzow, Ch. Duschek, G. Just, W. Pritzkow, and H. Schmidt, *J. Prakt. Chem.*, *317*, 53 (1975).

41 K. Okamoto, I. Nitta, T. Imoto, and H. Shingu, *Bull. Chem. Soc. Japan*, *40*, 1905 (1967).

42 C. K. Hancock, E. A. Meyers, and B. J. Yager, *J. Amer. Chem. Soc.*, *83*, 4211 (1961).

43 C. K. Hancock, and C. P. Falls, *J. Amer. Chem. Soc.*, *83*, 4214 (1961).

44 C. K. Hancock, B. J. Yager, C. P. Falls, and J. O. Schreck, *J. Amer. Chem. Soc.*, *85*, 1297 (1963).

45 M. S. Newman, *J. Amer. Chem. Soc.*, *72*, 4783 (1950).

46 O. Rosado-Lojo, C. K. Hancock, and A. Danti, *J. Org. Chem.*, *31*, 1899 (1966).

47 P. D. Bolton, *Austral. J. Chem.*, *19*, 1013 (1966).

48 Y. Inaki, S. Nozakura, and S. Murahashi, *Bull. Chem. Soc. Japan*, *42*, 2605 (1969).

49 J. P. Idoux, P.T.R. Hwang, and C. K. Hancock, *J. Org. Chem.*, *38*, 4239 (1973).

50 J. P. Idoux and J. A. Sikorski, *J. Chem. Soc. Perkin Trans. II*, 921 (1972).

51 G. Borkent and J. J. van Aartsen, *Rec. Trav. Chim.*, *91*, 1079 (1972).

52 I. Naito, A. Kinoshita, and T. Yonemitsu, *Bull. Chem. Soc. Japan*, *49*, 339 (1976).

53 B. Giese, *Angew. Chem. Internat. Edn.*, *15*, 688 (1976).

54 H.-D. Beckhaus and C. Rüchardt, *Chem. Ber.*, *110*, 878 (1977).

55 D. Grosjean, G. Mouvier, and J.-E. Dubois, *J. Org. Chem.*, *41*, 3872 (1976).

56 T. Fujita, *Pure and Appl. Chem.*, *50*, 987 (1978).

57 V. A. Palm, *Fundamentals of the Quantitative Theory of Organic Reactions*, (in Russian) Khimiya, Leningrad, 1967; revised edition, 1977. The first Russian edition was translated into German by G. Heublein as *Grundlagen der Quantitativen Theorie Organischer Reaktionen*, Akademie- Verlag, (East) Berlin, 1971.

58 *Reaktsionnaya Sposobnost Organicheskikh Soedinenii* (with English summaries), V. A. Palm, editor, Tartu, Esthonia, began in 1964. From 1966-1972 it was available in English translation as *Organic Reactivity*, Consultants Bureau, New York. Since 1974 it has been published from Tartu in separate Russian and English editions, the latter under the title *Organic Reactivity*.

59 I. A. Koppel, *Reakts. Sposobnost Org. Soedin.*, *2* (2), 26 (1965).

60 T. L. Brown, *J. Amer. Chem. Soc.*, *81*, 3229 (1959).

61 C. D. Ritchie and W. F. Sager, *Progr. Phys. Org. Chem.*, *2*, 323 (1964).

62 J. I. Brauman and L. K. Blair, *J. Amer. Chem. Soc.*, *91*, 2126 (1969).

63 M.S.B. Munson, *J. Amer. Chem. Soc.*, *87*, 2332 (1965).

64 F. W. Baker, R. C. Parish, and L. M. Stock, *J. Amer. Chem. Soc.*, *89*, 5677 (1967).

65 P. von R. Schleyer and C. W. Woodforth, *J. Amer. Chem. Soc.*, *90*, 6528 (1968).

66 L. M. Jackman and D. P. Kelly, *J. Chem. Soc. (B)*, 102 (1970).

67 P.M.E. Lewis and R. Robinson, *Tetrahedron Lett.*, 2783 (1970).

68 R. W. Taft, personal communication, cited as ref. 170 in ref. 1.

69 R. M. Noyes, *J. Amer. Chem. Soc.*, *86*, 971 (1964).

70 N. B. Chapman, J. Shorter, and K. J. Toyne, *J. Chem. Soc.*, 2543 (1961).

71 P. D. Bolton and G. L. Jackson, *Austral. J. Chem.*, *24*, 471 (1971).

72 P. D. Bolton and G. L. Jackson, *Austral. J. Chem.*, *24*, 969 (1971).

73 I. V. Talvik, *Reakts. Sposobnost Org. Soedin.*, 1 (2), 241 (1964).

74 J. A. MacPhee, A. Panaye, and J.-E. Dubois, *Tetrahedron*, *34*, 3553 (1978).

75 A. Panaye, J. A. MacPhee, and J.-E. Dubois, *Tetrahedron*, *36*, 759 (1980).

76 J.-E. Dubois, J. A. MacPhee, and A. Panaye, *Tetrahedron*, *36*, 919 (1980).

77 J. A. MacPhee, A. Panaye, and J.-E. Dubois, *J. Org. Chem.*, *45*, 1164 (1980).

78 J. P. Idoux and J. O. Schreck, *J. Org. Chem.*, *43*, 4002 (1978).

79 T. Fujita, C. Takayama, and M. Nakajima, *J. Org. Chem.*, *38*, 1623 (1973).

80 C. Takayama, T. Fujita, and M. Nakajima, *J. Org. Chem.*, *44*, 2871 (1979).

81 D. F. DeTar and C. J. Tenpas, *J. Amer. Chem. Soc.*, *98*, 4567 (1976).

82 D. F. DeTar and C. J. Tenpas, *J. Amer. Chem. Soc.*, *98*, 7903 (1976).

83 D. F. DeTar, D. F. McMullen, and N. P. Luthra, *J. Amer. Chem. Soc.*, *100*, 2484 (1978).

84 D. F. DeTar and N. P. Luthra, *J. Amer. Chem. Soc.*, *100*, 4505 (1980).

85 D. F. DeTar, *J. Org. Chem.*, *45*, 5174 (1980).

86 D. F. DeTar, *J. Org. Chem.*, *45*, 5166 (1980).

87 P. Müller and J.-C. Perlberger, *J. Amer. Chem. Soc.*, *98*, 8407 (1976).

88 J.-C. Perlberger and P. Müller, *J. Amer. Chem. Soc.*, *99*, 6316 (1977).

89 Ref. 4, p.599 *et seq.*

90 M. Charton, *J. Amer. Chem. Soc.*, *91*, 615 (1969).

91 M. Charton, *Progr. Phys. Org. Chem.*, *10*, 81 (1973).

92 M. Charton, *J. Amer. Chem. Soc.*, *97*, 1552 (1975).

93 M. Charton, *J. Org. Chem.*, *43*, 3995 (1978) is Part 13.

94 M. Charton, *J. Amer. Chem. Soc.*, *97*, 3691 (1975).

95 M. Charton, *J. Org. Chem.*, *41*, 2906 (1976).

96 M. Charton, *J. Amer. Chem. Soc.*, *99*, 5687 (1977).

97 M. Charton, *J. Org. Chem.*, *44*, 903 (1979).

98 A. J. MacPhee and J.-E. Dubois, *Tetrahedron Lett.*, 2471 (1976).

99 M. Charton, *J. Org. Chem.*, *29*, 1222 (1964).

100 F. G. Bordwell and H. E. Fried, *Tetrahedron Lett.*, 1121 (1977).

101 D. F. DeTar, *J. Amer. Chem. Soc.*, *102*, 7988 (1980).

102 L. S. Levitt and H. F. Widing, *Progr. Phys. Org. Chem.*, *12*, 119 (1976).

103 R. W. Taft and L. S. Levitt, *J. Org. Chem.*, *42*, 916 (1977).

104 R. W. Taft, M. Taagepera, J.L.M. Abboud, J. F. Wolf, D. J. DeFrees, W. J. Hehre, J. E. Bartmess, and R. T. McIver, *J. Amer. Chem. Soc.*, *100*, 7765 (1978).

105 J. E. Bartmess, J. A. Scott, and R. T. McIver, *J. Amer. Chem. Soc.*, *101*, 6056 (1979).

106 C. G. Screttas, *J. Org. Chem.*, *44*, 1471 (1979).

107 C. G. Screttas, *J. Org. Chem.*, *44*, 3332 (1979).

108 See also C. G. Screttas, *J. Org. Chem.*, *45*, 333 (1980).

109 C. Minot, O. Eisenstein, P. C. Hiberty, and N. T. Anh, *Bull. Soc. Chim. France II*, 119 (1980).

110 W. Adcock and T.-C. Khor, *J. Org. Chem.*, *43*, 1272 (1978).

111 A recent example is provided by: L. M. Sayre and F. R. Jensen, *J. Amer. Chem. Soc.*, *101*, 6001 (1979).

112 M. Charton, *Progr. Phys. Org. Chem.*, *13*, 119 (1981).

113 C. A. Grob and M. G. Schlageter, *Helv. Chim. Acta*, *59*, 264 (1976).

114 Part VII, C. A. Grob and R. Rich, *Helv. Chim. Acta*, *62*, 2793 (1979).

115 A. G. Burden, N. B. Chapman, J. Shorter, K. J. Toyne, and L. M. Wilson, *J. Chem. Soc. Perkin Trans. II*, 1212 (1980).

116 M. Sjöström and S. Wold, *J. Chem. Soc. Perkin Trans. II*, 1274 (1979).

117 General account of substituent effects in C. K. Ingold, *Structure and Mechanism in Organic Chemistry*, Bell, London, 2nd edn., 1969.

118 H.B. Watson, *Modern Theories of Organic Chemistry*, Oxford U.P., 2nd edn., 1941, p.241.

119 W. F. Little, C. N. Reilley, J. D. Johnson, K. N. Lynn, and A. P. Sanders, *J. Amer. Chem. Soc.*, *86*, 1376 (1964).

120 A. F. Hegarty and F. L. Scott, *J. Org. Chem.*, *32*, 1957 (1967).

121 K. Kindler, *Ann. Chem.*, *464*, 278 (1928).

122 C. Eaborn, D.R.M. Walton, and D. J. Young, *J. Chem. Soc. (B)*, 15 (1969).

123 J. G. Watkinson, W. Watson, and B. L. Yates, *J. Chem. Soc.*, 5437 (1963).

124 D. H. McDaniel and H. C. Brown, *J. Amer. Chem. Soc.*, *77*, 3756 (1955).

125 I. J. Solomon and R. Filler, *J. Amer. Chem. Soc.*, *85*, 3492 (1963).

126 For example, M. T. Tribble and J. G. Traynham, *J. Amer. Chem. Soc.*, *91*, 379 (1969).

127 M. Charton, *J. Amer. Chem. Soc.*, *91*, 6649 (1969).

128 M. Charton, *Progr. Phys. Org. Chem.*, *8*, 235 (1971); detailed references in ref. 1.

129 M. H. Aslam, A. G. Burden, N. B. Chapman, J. Shorter, and M. Charton, *J. Chem. Soc. Perkin Trans. II*, 500 (1981).

130 M. H. Aslam, N. B. Chapman, J. Shorter, and M. Charton, *J. Chem. Soc. Perkin Trans. II*, 720 (1981).

131 N. B. Chapman, J. Shorter, and J.H.P. Utley, *J. Chem. Soc.*, 1824 (1962).

132 A. Buckley, N. B. Chapman, and J. Shorter, *J. Chem. Soc. (B)*, 195 (1969).

133 M. Charton, *J. Org. Chem.*, *40*, 407 (1975).

134 H. Mager, P. P. Mager, and A. Barth, *Tetrahedron*, *35*, 1953 (1979).

135 H. Mager, *Tetrahedron*, *37*, 509 (1981).

136 P. P. Mager, *Sci. Pharm.*, *49*, 1 (1981).

137 H. Mager, *Tetrahedron*, *37*, 523 (1981).

138 M. Godfrey, *Tetrahedron Lett.*, 3203 (1972).

139 T. M. Krygowski and W. R. Fawcett, *J. Chem. Soc. Perkin Trans. II*, 2033 (1977).

140 T. Fujita and T. Nishioka, *Progr. Phys. Org. Chem.*, *12*, 49 (1976).

141 T. Nishioka, T. Fujita, K. Kitamura, and M. Nakajima, *J. Org. Chem.*, *40*, 2520 (1975).

142 G. Consiglio, R. Noto, D. Spinelli, and C. Arnone, *J. Chem. Soc. Perkin Trans. II*, 219 (1979).

143 D. Spinelli, R. Noto, and G. Consiglio, *J. Chem. Soc. Perkin Trans. II*, 747 (1976).

144 D. Spinelli, R. Noto, G. Consiglio, and A. Storace, *J. Chem. Soc. Perkin Trans. II*, 1805 (1976).

145 L. L. McCoy and E. E. Riecke, *J. Amer. Chem. Soc.*, *95*, 7407 (1973).

146 R. N. McDonald, R. R. Reitz, and J. M. Richmond, *J. Org. Chem.*, *41*, 1822 (1976).

147 U. Berg, R. Gallo, G. Klatte, and J. Metzger, *J. Chem. Soc. Perkin Trans. II*, 1350 (1980).

148 S. Berger, *Tetrahedron*, *32*, 2451 (1976).

149 C. H. Yoder, F. K. Sheffy, R. Howell, R. E. Hess, L. Pacala, C. D. Schaeffer, and J. J. Zuckerman, *J. Org. Chem.*, *41*, 1511 (1976).

CHAPTER 5
Solvent Effects

Page

5.1. Introduction 128
5.1.1. Qualitative theory of the influence of the solvent on reaction rate 130
5.1.2. Treatment of solvent effects on reaction rate in terms of dielectric constant 132
5.1.3. The limitations of dielectric constant 136

5.2. Solvent parameters 138
5.2.1. The Grunwald and Winstein Y and related parameters 139
5.2.2. Other 'chemical' solvent parameters 142
5.2.3. The Z-values 142
5.2.4. E_{T} values of N-phenol-pyridinium betaines 143
5.2.5. Brownstein's LFER 144

5.3. The concept of solvent polarity 145

5.4. The Koppel—Palm treatment 146
5.4.1. Introduction and applications 146
5.4.2. Comments on the Koppel—Palm treatment 154
5.4.3. Alternative solvent parameters 158
5.4.4. The Kamlet—Taft treatment 160

5.5. Miscellaneous topics 165
5.5.1. The correlation analysis of solvent effects by means of substituent constants 165
5.5.2. Solvent effects on ρ and σ values 167
5.6. References 170

5.1. INTRODUCTION

The effect of different solvents upon rates of chemical change was one of the earliest kinetic problems to be studied. The first studies were done more than a century ago. However the development of correlation analysis in the area of solvent effects has proved to be a slow and difficult process and only within the last 20 years has there been any considerable progress. The application of the techniques of multiple regression has now proved strikingly successful and has greatly increased our understanding of the role of the solvent. To appreciate the present position, it is necessary to know something of the way the study of solvent effects has developed.

Berthelot and Saint-Gilles were the first to observe a solvent effect on organic reactivity, in their studies of the esterification of acetic acid by ethanol in 1862.[1] The first systematic studies of solvent effects, however, were made by Menschutkin in St. Petersburg between about 1887 and 1900. Menschutkin's earliest work was on the reaction between acetic anhydride and an alcohol,[2] but he is best known for his work on the rates of formation of quaternary ammonium halides from tertiary amines and alkyl halides.[3] For the reaction between triethylamine and ethyl iodide he measured rate constants in more than 20 varied solvents. The reaction was over 740 times faster in benzyl alcohol, the fastest solvent, than in hexane, the slowest. His results convinced him that "a reaction cannot be separated from the medium in which it takes place". He was unable to give any detailed interpretation of his results, but it seemed to him that the chemical nature of the solvent was more important than its physical properties in governing the rate of reaction.

The first thorough-going work on the influence of solvents on equilibrium constants was done about 1896 on keto-enol and nitro-isonitro tautomerism, and involved such distinguished names as Claisen,[4] Wislicenus,[5] Knorr,[6] and Hantzsch.[7] We shall not be much concerned with solvent effects on equilibria. Indeed apart from solvent effects on acid—base equilibria, such studies have played a relatively minor role and the study of solvent effects on organic reactivity usually involves rate constants rather than equilibrium constants.

Between 1900 and 1930 there were many studies of solvent effects on rates, and attempts were made to relate them to various solvent properties. With hindsight we can, of course, see that even a superficial understanding could not precede considerable knowledge of reaction mechanisms, molecular structure, the transition state, and intermolecular forces. For an example of work in this period see Patterson and Montgomerie.[8]

The 1920s saw the remarkable success of the application of collision theory by Hinshelwood and others to be the kinetics of gas reactions.[9] In the 1930s attempts were made to extend this treatment to the kinetics of reactions in solution. This was the main purpose of Moelwyn-Hughes's book *The Kinetics of Reactions in Solution* published in 1933.[10] In this work the initial aim was to identify reactions for which a simple collisional interpretation of the Arrhenius equation appeared applicable, and then to examine and account for systems which did not conform to the simple treatment. This involved the role of the solvent, and Moelwyn-Hughes devoted about 13 pages to discussing this for so-called 'slow' bimolecular reactions, mainly of the Menschutkin type.[11] The discussion includes three pages on "The Attempted Correlation of Velocity Constants with Certain Physical Properties", these being dielectric constant, cohesion, and viscosity.

In the early 1930s transition state theory was developed and within a short time physical chemists were applying this to the kinetics of reactions in solution. In particular there were the contributions of Laidler and Eyring, who developed the quantitative treatment of solvent effects on rates in terms of models of the initial state and the activated complex.[12] The principal forces operating were considered to be electrostatic interactions involving ions and/or dipoles as appropriate, such interactions being moderated by the dielectric behaviour of the solvent. In this work we may see the rudimentary beginnings of the correlation analysis of solvent effects, in so far as it involves a linear relationship of log k to a solvent property. We will refer to this work again later. First, however, we must note that parallel with these developments was the work of the organic chemists, who attempted a more qualitative approach, based on classifying reactions according to 'charge-type' and considering the interaction between solvent and reactants in the initial state and in the transition state for the various cases. The work of Hughes and Ingold from about 1933 onwards thoroughly laid the foundations of this treatment but tentative moves on the same lines had been made slightly earlier, *e.g.* by Richardson and Soper,[13] and by Soper and Williams.[14]

5.1.1. QUALITATIVE THEORY OF THE INFLUENCE OF THE SOLVENT ON REACTION RATE

The first paper by Hughes and Ingold was in 1935.[15] For a full account see Ingold's book.[16]

The basic ideas may be stated as follows:

(*a*) The rates of reactions in which ionic charges arise or are compressed into a smaller space during passage of the reactants from the initial state to the transition state increase with increase of the polarity of the medium surrounding the reactants and the activated complex.

(*b*) Conversely if ionic charges disappear or are dispersed throughout a larger space during the passage to the transition state, the reaction is retarded by an increase in the polarity of the medium.

(*c*) Reactions in which ionic charges arise or disappear are much more strongly influenced by the polarity of the medium than those in which the activation process simply involves charge diffusion.

The use of the term 'polarity' in these statements is now conventional but the word does not in fact occur in Hughes and Ingold's paper in 1935.[15] There they refer to 'strongly solvating' or 'feebly solvating' solvents rather than polar or non-polar solvents. However, when Ingold presented this work in 1953 for the first edition of *Structure and Mechanism in Organic Chemistry* he used 'solvent polarity' and related terms.[16] An extended quotation is appropriate.[17] He wrote:

"Concerning solvents, it has been assumed that polarity, that is power to solvate charges in solutes, will

(1) increase with the molecular dipole of the solvent;

(2) decrease with increased thickness of shielding of the dipole charges.

The hydroxyl group thus becomes a strong factor in solvent polarity, because the proton is so thinly shielded; and water, the lower alcohols, aprotic dipolar solvents, and non-dipolar solvents, for example, H_2O, EtOH, Me_2CO, C_6H_6 form a series of solvents in order of diminishing polarity."

Item (2) and the comment about it are of interest in relation to our later discussion (Section 5.3) of the concept of solvent polarity.

The above statement of the basic ideas indicates that the charge-type to which the reaction belongs is of fundamental importance, and Hughes and Ingold developed a systematic classification of reactions in terms of it. Charge-type governs the extent of interaction between the solvent and the reactants in the initial state and in the transition state. When the passage to the transition state involves the creation of ionic charges or a concentration of charge, the activated complex will be more strongly solvated than the original reactants. An increase in the polarity of the solvent will therefore favour the transition state relative to the initial state. Conversely, when the activation process involves the disappearance of ionic charges or a dispersal of charge, the reactants will be solvated more strongly in the initial state than in the transition state. An increase in the polarity of the solvent will then favour the initial state relative to the transition state.

We now illustrate the treatment by two well-known examples.

The solvolysis of t-butyl chloride in hydroxylic solvents proceeds by the S_N1 mechanism, in which the rate-determining step is the heterolysis of the C—Cl bond:

$$Me_3CCl \rightleftharpoons [Me_3\overset{\delta+}{C}\text{--}\overset{\delta-}{Cl}]^{\ddagger} \rightleftharpoons Me_3\overset{+}{C} + Cl^- \xrightarrow[\text{with solvent}]{\text{Reaction}} \text{products} \qquad (5.1)$$

This is a reaction in which ionic charges arise during the

formation of the activated complex. It therefore proceeds much more rapidly in water, which is a very polar solvent, than in the alkanols, which are much less polar. The relative rate constant as between water and ethanol is about 10^5.

Reactions between nucleophilic ions Y^- and primary alkyl halides RX proceed by the S_N2 mechanism, in which there is synchronous breaking of the C—X bond and formation of the C—Y bond:

$$Y^- + RX \rightleftharpoons [\overset{\delta_1^-}{Y}\text{--}R\text{--}\overset{\delta_2^-}{X}]^{\ddagger} \rightleftharpoons YR + X^- \qquad (5.2)$$

This is a type of reaction in which charge is dispersed throughout a larger space during the activation process. It is therefore retarded by an increase in the polarity of the solvent, but the change on going from ethanol to water is not so dramatic as in the previous example: depending on the nucleophile and the substrate, retardation may be by a factor of up to about 50.

The Hughes and Ingold treatment of solvent effects is certainly an over-simplified approach, particularly in its dealing with aqueous organic solvent mixtures, and in assuming the dominance of enthalpy rather than entropy effects. However, it works well over a wide area, and its emphasis on the importance of charge distribution in reactants and in the activated complex is undoubtedly correct.

5.1.2. TREATMENT OF SOLVENT EFFECTS ON REACTION RATE IN TERMS OF DIELECTRIC CONSTANT

For this purpose bimolecular reactions may be classified as dipole—dipole, ion—dipole or ion—ion reactions. The last-mentioned category is not common in organic chemistry; we will deal in turn with the first two types (See Laidler, 1963 for details.[18])

Dipole—Dipole Reactions

Kirkwood[19] derived expression (5.3) for the change in the free energy of a molecule of dipole moment μ and radius r when transferred from a medium of unit dielectric constant to one of dielectric constant ε.

$$\Delta G = -(\mu^2/r^3)(\varepsilon - 1)/(2\varepsilon + 1) \qquad (5.3)$$

The application of (5.3) and the appropriate equation of transition state theory (equation 1.2) to a reaction between dipolar molecules A and B gives equation (5.4), where ‡ designates the parameters of the activated complex and k' is the rate constant for a medium of unit dielectric constant. (It is assumed that only electrostatic forces need be considered.)

$$\log k = \log k' - \frac{1}{2.303\,\mathrm{k}T} \cdot \frac{\varepsilon - 1}{2\varepsilon + 1}\left[\frac{\mu_A^2}{r_A^3} + \frac{\mu_B^2}{r_B^3} - \frac{\mu_\ddagger^2}{r_\ddagger^3}\right] \quad (5.4)$$

The formation of the activated complex is often accompanied by considerable charge separation, in which case $\mu_\ddagger$ is very much greater than μ_A or μ_B. Equation (5.4) then predicts that the plot of log k against $(\varepsilon - 1)/(2\varepsilon + 1)$ will be a straight line of positive slope. If values for all the quantities in square brackets could be suggested, equation (5.4) would predict the actual slope. Most of the quantities are somewhat ill-defined, particularly those for the transition state, and the best that can be done quantitatively is to work backwards and show that the observed slope fits in with 'reasonable' values.

For many reactions in binary solvent mixtures, whose dielectric constants change with composition, good linear plots of log k versus $(\varepsilon - 1)/(2\varepsilon + 1)$ are obtained. This applies, for example, to the Menschutkin reaction (5.5) in mixtures of acetone with dioxan, benzene, or tetrahydrofuran.

$$Et_3N \quad + \quad EtI \quad \rightarrow \quad Et_4N^+ \quad + \quad I^- \quad (5.5)$$

However, each binary solvent system leads to its own straight line of distinctive slope, which is not at all in accord with equation (5.4). This clearly indicates the limitations of dielectric constant and the importance of specific interactions between solvent and the various solutes. Presumably the Kirkwood function of dielectric constant 'works' for each binary solvent system because the solutes are effectively saturated with the specific interactions over the composition range employed. Sometimes, however, curvilinear relationships of log k to the Kirkwood function are found.

For the same reaction in a series of one-component solvents there is a general tendency for log k to increase with $(\varepsilon - 1)/(2\varepsilon + 1)$, but there are many anomalies; see Fig. 5.1. Notably the points for hexane and cyclohexane deviate markedly.

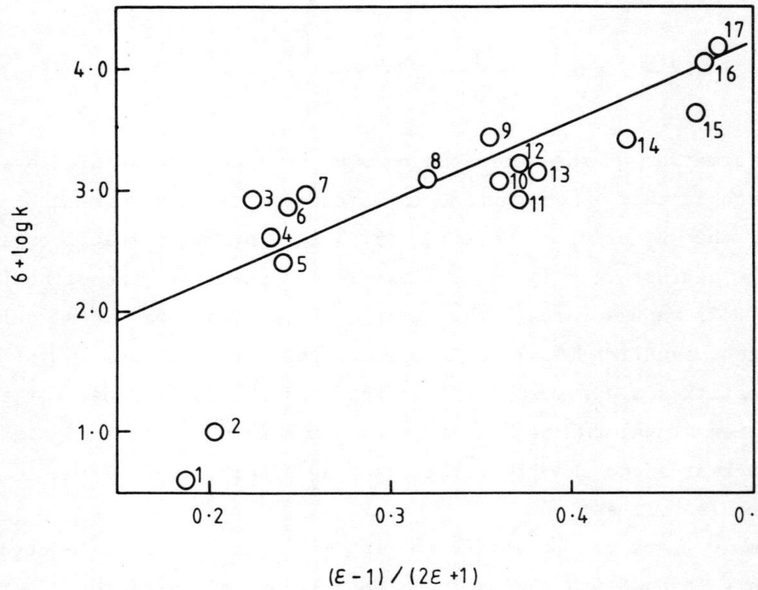

FIG. 5.1. Attempted correlation of log k ($l\ mol^{-1}\ sec^{-1}$, 100^oC) for reaction between triethylamine and ethyl iodide with the Kirkwood function of dielectric constant. After J. Shorter, *Correlation Analysis in Organic Chemistry*, Oxford Chemistry Series, Clarendon Press, Oxford, 1973, p.68. Key to solvents: 1. hexane; 2. cyclohexane; 3. dioxan; 4. benzene; 5. toluene; 6. *p*-dichlorobenzene; 7. diphenylmethane; 8. diphenyl ether; 9. iodobenzene; 10. *m*-dichlorobenzene; 11. fluorobenzene; 12. bromobenzene; 13. chlorobenzene; 14. *o*-dichlorobenzene; 15. acetone; 16. benzonitrile; 17. nitrobenzene.

The moderate success of the correlation for most of the points is probably due to most of the solvents being substituted benzenes, but no doubt even among these, specific solvent—solute interactions vary quite widely. Also the *polarizability* of the solvent molecule may be important, *e.g.* with the halogenobenzenes the reaction is fastest in the member of lowest dielectric constant, iodobenzene, the most polarizable of the halogenobenzenes.

Data for a very wide range of solvents are available for the closely related reaction of tri-n-propylamine and methyl iodide (see Fig. 5.2). Correlation of log k with $(\varepsilon - 1)/(2\varepsilon + 1)$ is now

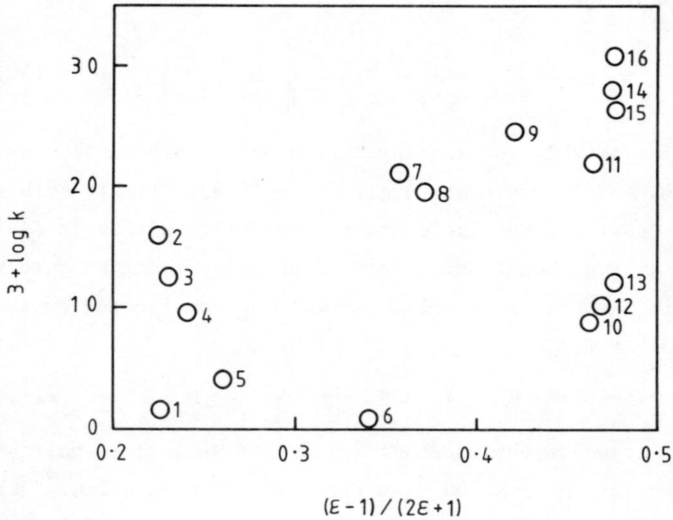

FIG. 5.2. Attempted correlation of log k (l mol^{-1} min^{-1}, 20°C) for reaction between tri-n-propylamine and methyl iodide with the Kirkwood function of dielectric constant. After J. Shorter, *UV Spectrometry Group Bulletin*, (4), 37 (1976). Data from C. Lassau and J.-C. Jungers, *Bull. Soc. Chim. France*, 2678 (1968). Key to solvents: 1. carbon tetrachloride; 2. dioxan; 3. benzene; 4. toluene; 5. carbon disulphide; 6. diethyl ether; 7. chloroform; 8. bromobenzene; 9. methylene chloride; 10. n-propyl alcohol; 11. acetone; 12. ethanol; 13. methanol; 14. *NN*-dimethylformamide; 15. methyl cyanide; 16. nitromethane.

seen to be tenuous in the extreme.

Ion-Dipole Reactions

For an ion the equation corresponding to (5.3) for a dipole is (5.6), where Ze is the charge and r is the radius of the ion. The

$$\Delta G = -\frac{Z^2e^2}{r}\left[\frac{\varepsilon - 1}{2\varepsilon}\right] \qquad (5.6)$$

application of (5.6) and transition state theory to an ion—dipole reaction gives (5.7). The electrostatic free-energy contribution of the reactant dipole is neglected. Since $r_\ddagger > r$, a linear relationship between log k and $(\varepsilon - 1)/2\varepsilon$ of negative slope is predicted.

$$\log k = \log k' + \frac{Z^2e^2}{2.303kT}\left[\frac{\varepsilon - 1}{2\varepsilon}\right]\left[\frac{1}{r_\ddagger} - \frac{1}{r}\right] \qquad (5.7)$$

The data available for testing this equation are mainly for reactions in aqueous organic mixtures. The linear relationship is sometimes observed, but there are various difficulties in dealing with this kind of reaction. For instance any reaction involving OH^- ions in alcohol—water mixtures will be complicated by the equilibrium (5.8):

$$OH^- + ROH \rightleftharpoons OR^- + H_2O \qquad (5.8)$$

In certain solvents the reactant ion may be subject to pairing with its counter ion, *e.g.* Na^+OH^-, particularly when ε is low. Part of the observed solvent effect may arise because the ion pair is much less reactive than the free anion.

5.1.3. THE LIMITATIONS OF DIELECTRIC CONSTANT

The application of 'bulk' dielectric constant to interactions on a molecular scale is, in principle, inadmissible. The solvent cannot be 'between' the ions or dipoles, which are about to react, in the same sense that the solvent is between the plates of a condenser in the measurement of ε. For binary solvent mixtures the objection is even stronger: solvent—solute interactions may tend to concentrate one component in the vicinity of the molecules of one of the reactants. The local composition of the solvent will then not be the same as the overall composition, and the local dielectric constant cannot possibly be the same as that corresponding to the given overall composition. Nevertheless, as we have seen, the application of ε is attended with some success, particularly (and paradoxically!) in the realm of mixed solvents. We shall see later how more refined treatments of solvent effects, involving multiple regression, often retain ε as a measure of the influence of the solvent on general electrostatic interactions on the molecular scale.

It is sometimes suggested that ε should be replaced by n_D^2 as a measure of 'dynamic' dielectric constant; n_D^2 is the square of the refractive index for sodium light.[20] (n_D^2 rather than n_D is used because of the connection between the dielectric constant of a medium and the square of its refractive index which is part of J. Clerk Maxwell's electromagnetic theory of light. Values for sodium light are used because these are available for numerous liquids.) Dielectric constants are measured in a low frequency field in which dipoles can be orientated to follow the field; refractive index involves interaction with a high frequency field, and is a measure of *polarizability*, whereas dielectric constant includes *polarization*. The reason for replacing ε by n_D^2 is that the rapidity of molecular processes might make n_D^2 more relevant to solvent effects than ε. Equations (5.4) and (5.7) can be re-written with n_D^2 replacing ε. The amended equations are usually less successful than the original, but n_D^2 may be used to supplement ε (see Section 5.4.1).

The strongest criticism of the use of ε in connection with solvent—solute interactions is that it does not express the specific nature of such interactions. Solvent molecules may have highly localized centres which interact specifically with localized centres in solutes. Such interactions are not adequately represented by radius, overall charge, and dipole moment for solutes, with moderation by bulk dielectric constant of solvent. A protic solvent (alcohol, carboxylic acid, *etc.*) may be able to form hydrogen bonds with suitable negative centres in the solute. Also any solvent with unshared electron pairs may enter into donor—acceptor interaction with electron-poor centres in the solute. This is particularly important for dipolar aprotic solvents, *e.g.* acetone, dimethyl sulphoxide, and *NN*-dimethylformamide.

We now give two examples which show the importance of hydrogen bonding. The S_N2 reaction between aniline and ω-bromoacetophenone:

$$PhNH_2 + PhCO{\cdot}CH_2Br \rightarrow Ph\overset{+}{N}H_2{\cdot}CH_2{\cdot}COPh + Br^- \qquad (5.9)$$

involves the formation of a highly polar activated complex. Protic solvents interact very strongly with the $Br^{\delta-}$ end of the activated complex through hydrogen bonding, thereby stabilizing the complex and facilitating reaction. Thus the rate constant for methanol is about six times that for nitrobenzene, although the dielectric constants of the solvents are almost the same.

Anion—dipole reactions are much faster in dipolar aprotic solvents than in alcohols. Thus the reaction between *p*-fluoronitrobenzene and azide ion:

$$p\text{-}NO_2C_6H_4F + N_3^- \rightarrow p\text{-}NO_2C_6H_4N_3 + F^- \qquad (5.10)$$

is much faster in DMF than in methanol (by a factor $> 10^4$) even though the dielectric constants differ only slightly. The azide ion is heavily solvated by hydrogen bonding in methanol, and much of the solvation shell must be removed in forming the activated complex. In DMF the azide ion is in a much more exposed and therefore reactive condition.

The recognition of the limitations of dielectric constant has led various authors to attempt to dispense with it entirely in favour of *solvent parameters* based on actual solvent-sensitive chemical or physical processes. In this way the originators of each solvent parameter hope to allow for the complexity of solvent—solute interactions. We shall now examine some of the best-known solvent parameters [see Reichardt (1979),[21] Griffiths and Pugh (1979),[22] and Koppel and Palm (1972)[23]].

5.2. SOLVENT PARAMETERS

The majority of these are based either on linear free-energy relationships for chemical processes or on solvatochromic[†] shifts in electronic spectra. We consider, first, examples in the former category.

5.2.1. THE GRUNWALD AND WINSTEIN Y AND RELATED PARAMETERS

These authors (1948)[24] suggested treating solvent effects on rates in terms of equation (5.11), similar in form to the Hammett equation.

$$\log k = \log k^{\text{o}} + m\text{Y} \qquad (5.11)$$

Log k refers to a given reaction in a given solvent, log k^{o} to the same reaction in 80% v/v aqueous ethanol as a standard solvent, Y is a parameter characteristic of the given solvent and m is a parameter characteristic of the given reaction, which measures its susceptibility to changes in solvent; the analogy of Y and m to σ and ρ respectively is obvious. Scales of Y and m were established by taking Y = 0 for 80% ethanol, and selecting the solvolysis of t-butyl chloride at 25°C as a standard reaction, for which m is defined as 1.000. Y values are known for various one-component solvents (mainly alcohols) and for various mixtures of organic solvents with water or a second organic solvent. Selected values of Y are in Table 5.1. From the nature of the standard reaction, Y values will clearly provide some measure of the ability of the solvent to separate ions. It was therefore hoped that Y values might be a more generally applicable measure of solvent polarity than dielectric constants.

In a large number of cases the Grunwald—Winstein equation is fairly successful, and in some cases log k appears to be correlated better with Y than with functions of dielectric constant. Good linear relationships between log k and Y are shown by the solvolyses of various halides and secondary sulphonates, *i.e.* reactions which proceed by an S_N1 mechanism, like the standard reaction. The situation for S_N2 reactions (*e.g.* primary halide solvolysis) or for reactions of borderline mechanism (*e.g.* secondary halide solvolysis)

† Solvatochromic shift and solvatochromism mean the displacement (to higher or lower frequency) of a spectral band, as caused by interaction of the absorbing molecule with molecules of solvent.
Footnote to p. 138

TABLE 5.1. Solvent parameters[a]

Solvent	ε^{b}	Y^{c}	Z^{c}	$E_T{}^{c}$
Cyclohexane	2.02	-	60.1	31.2
Dioxan	2.21	-	64.6	36.0
Carbon tetrachloride	2.23	-	-	32.5
Benzene	2.28	-	54.0	34.5
Chloroform	4.81	-	63.2	39.1
Acetic acid	6.2	-1.639	79.2	51.2[d]
Dichloromethane	8.9[c]	-	64.2	41.1
Pyridine	12.3[c]	-	64.0	40.2
t-Butyl alcohol	12.5[c]	-3.26	71.3	43.9[e]
i-Propyl alcohol	18.3[c]	-2.73	76.3	48.6
Acetone	20.7[c]	-	65.7	42.2
Ethanol	24.3[c]	-2.033	79.6	51.9
Methanol	33.6	-1.09	83.6	55.5
Nitrobenzene	34.8[c]	-	-	42.0
Acetonitrile	37.5	-	71.3	46.0
NN-Dimethylformamide	38.3	-	68.5	43.8
Dimethyl sulphoxide	46.6[c]	-	70.2	45.0
Water	80.4	3.493	94.6[f]	63.1
Formamide	111.0	0.604	83.3	56.6
Ethanol: water, 80:20	34.2[c]	0.000	84.8	53.6
Acetone: water, 80:20	33.6[c]	-0.673	80.7	52.2

a Assembled mainly from compilations in refs. 21-23.

b At 20°C, except as indicated.

c At 25°C.

d Based on a correlation between E_T and Z.

e At 30°C.

f Based on a correlation between Y and Z.

is less satisfactory. Also, if log k is plotted against Y for the solvolysis of diphenylmethyl chloride in aqueous ethanol, aqueous acetone, or aqueous dioxan, each binary solvent system gives a different line of characteristic slope. If the Grunwald—Winstein equation were strictly obeyed all the points should lie on a single straight line.

The above limitations of the Grunwald—Winstein equation may indicate that the rates of reaction depend not only on the 'ionizing power' of the solvent (considered to be measured by Y) but also on the 'nucleophilicity' of the solvent. Such a property may clearly be relevant for solvolyses with substantial S_N2 characteristics. These considerations have led to attempts to improve the Grunwald—Winstein equation. This is a fairly complicated story, which is well summarized by Reichardt.[25] Briefly we may note that the status of the Y value as originally defined by Grunwald and Winstein has been exhaustively confirmed, *i.e.* it is a fairly pure measure of 'ionizing power', unadulterated by solvent nucleophilicity. The Grunwald—Winstein equation has been generalized in the form:[26]

$$\log k = \log k^{o} + mY + lN \qquad (5.12)$$

where m and l are substrate parameters, Y is the solvent ionizing power and N measures solvent nucleophilicity. A scale of N has been developed on the basis that $l = 1.00$ and $m = 0.3$ for the solvolysis of methyl tosylate.[27,28] For the solvolyses of various primary and secondary substrates equation (5.12) gives significantly improved correlations compared with the simple equation (5.11). It is possible that the equation may be further extended to include consideration of solvent electrophilicity. The latter characteristic was actually incorporated along with solvent nucleophilicity in an analogous correlation equation put forward by Swain, Mosely, and Bown in 1955.[29]

$$\log k = \log k^{o} + c_1 d_1 + c_2 d_2 \qquad (5.13)$$

The k^{o} is the rate coefficient in the standard solvent (80% aqueous ethanol), c_1 and c_2 measure the sensitivity of the substrate to nucleophilic and electrophilic solvent character respectively, and d_1 and d_2 are measures of the nucleophilic and electrophilic character of the solvent. The approach was essentially statistical, *i.e.* values of the *four* parameters c_1, c_2, d_1 and d_2 were calculated to give the best account of a large amount of data. Some of the values of the parameters which were calculated are, however, chemically absurd, *e.g.* t-butyl chloride has a higher c_1 value than methyl bromide, suggesting that methyl bromide (S_N2 mechanism) is *less* sensitive to solvent nucleophilicity than t-butyl chloride (S_N1 mechanism).[30]

5.2.2. OTHER 'CHEMICAL' SOLVENT PARAMETERS

Gielen and Nasielski[31] suggested that a polarity scale (X values) could be based on electrophilic aliphatic substitution, *cf.* Y and nucleophilic substitution. Doubtless X values are some measure of the ionizing power of the solvent, but again more specific effects probably play a part.

Among other chemical measures of solvent polarity which have been suggested are the log k values for the Menschutkin reaction of tri-n-propylamine and methyl iodide (20°).[32]

5.2.3. THE *Z*-VALUES

Solvatochromic behaviour has long been known but the idea that it could be used to characterize solvent polarity for the correlation analysis of solvent effects on reactivity appears to stem from the work of Kosower in 1958[33] (see also Kosower, 1968[34]). The charge-transfer absorption of 1-ethyl-4-methoxycarbonylpyridinium iodide shows marked solvatochromism, since the more polar solvents stabilize the ground state, which is an ion pair, relative to the excited state, which is a radical pair.

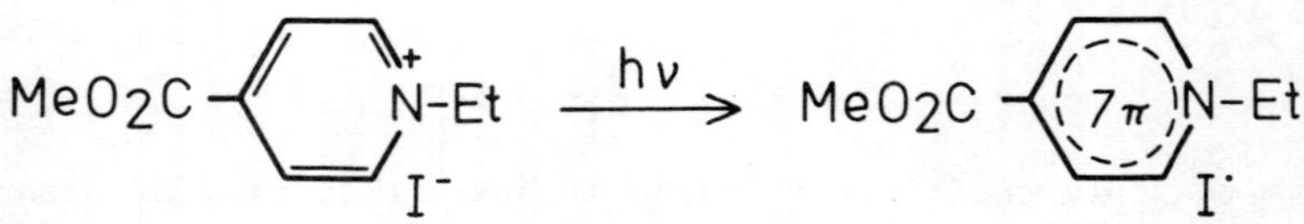

(5.14)

The *Z* value for a solvent is defined as the energy of this transition (kcal mol^{-1}) at $25^{\circ}C$, a high *Z* value corresponding to high solvent polarity. *Z* values cover the range 95 (water) to about 60 (iso-octane), and are available for a large number of one component solvents and mixtures. Selected values are in Table 5.1. Z gives excellent correlations with other solvatochromic parameters and with certain rate data, *e.g.* Menschutkin reactions. For some mixed solvents Z is related linearly to Y.

The last two decades have seen a proliferation of spectroscopically based solvent parameters: uv/visible/ir/nmr spectra have all been variously involved. At this stage we will mention only one more spectroscopic parameter, the E_T value originated about 1963 by Dimroth *et al.*[35]

5.2.4. E_T VALUES OF *N*-PHENOL-PYRIDINIUM BETAINES

The solvatochromic effect for the compound (I) with R = H is very large. The long-wave band is at 810 nm in diphenyl ether (transition energy in kcal mol^{-1}, E_T = 35.3) and at 453 nm in water (E_T = 63.1). The more polar solvents stabilize the zwitter-ionic ground state relative to the excited state.

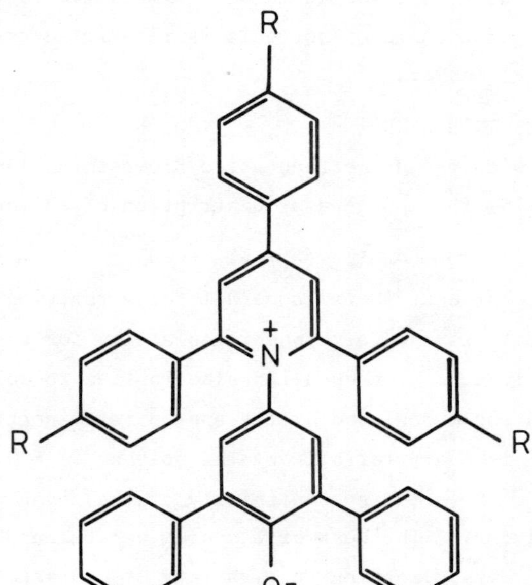

(I)

The compound with R = H is insoluble in hydrocarbons but the E_T scale can be extended to these by using the R = Me compound. The polarity scale based on compound (I) is strictly speaking the $E_T(30)$ scale, the compound being the thirtieth compound which the first authors examined. However the suffix (30) is now often dropped, it being understood that the usual E_T values are those based on the R = H compound. Selected values of E_T are in Table 5.1. More extensive compilations are in Reichardt's publications.[21,36] Values are available for certain mixed solvents but not for acidic solvents which protonate (I) at the O^-.

Certain solvents, *e.g.* formamide and dimethyl sulphoxide (DMSO), which are commonly regarded as highly polar (they have high dielectric constants and they readily dissolve salts) are by no means so polar in their solvatochromic behaviour; the E_T values of DMSO and t-butyl alcohol are comparable. No doubt the ability of protic solvents to form hydrogen bonds with the O^- in (I) is important.

E_T values have been used for correlating various kinetic and spectroscopic data. Quite good correlations are sometimes observed, although often the range of solvents is restricted and there are 'badly-behaved' members.

5.2.5. BROWNSTEIN'S LFER

By analogy with the Hammett equation Brownstein (1960)[37] suggested (5.15) for the general description of solvent effects:

$$\log (k_S/k_E) = SR \qquad (5.15)$$

k_S is the rate or equilibrium constant for a reaction in various solvents, and k_E is the corresponding quantity for absolute ethanol as solvent. The equation also applies to solvatochromic effects, with log k replaced by the appropriate spectroscopic quantity. S is characteristic of the solvent (S = 0.00 for ethanol) and R gives the susceptibility of the given property to change of solvent. The Kosower Z system was taken as standard (R = 1.00). By application to a variety of processes, 158 S values and 78 R values were derived.

This was an interesting attempt at wide generalization but many of the correlations are rather poor. No doubt too many different effects are being mixed-up and treated in an oversimplified way.

5.3. THE CONCEPT OF 'SOLVENT POLARITY'

It is now time to make plain the inadequacy of the concept of 'solvent polarity'. As a qualitative idea applied to a limited number of solvents the concept seems unobjectionable and useful: it relates to facilitating the separation of electric charges of unlike sign or the approach of charges of like sign. All chemists will agree that hexane is a non-polar solvent and water is highly polar, with ethanol somewhere in between. Difficulties start when we try to define solvent polarity more precisely and get worse when we try to express it quantitatively. It is one of those terms for which Alice's comment on the poem about the Jabberwock seems appropriate: "Somehow it seems to fill my head with ideas - only I don't exactly know what they are!!"[38a] In consequence, the remarks of another character in *Through the Looking-Glass* are also appropriate: "When I use a word", Humpty Dumpty said, in a rather scornful tone, "It means just what I choose it to mean — neither more nor less".[38b]

This situation regarding 'solvent polarity' has recently been recognized by no less a body than the International Union of Pure and Applied Chemistry.[39] In a glossary of terms used in physical organic chemistry, there is an entry under *Polarity* which reads: "When applied to a solvent, this rather ill-defined term is synonymous with *ionizing power*". Under the latter we may read: "A qualitative term to denote the tendency of a particular solvent to promote ionization of an uncharged solute".

Actually, as long ago as 1927, Sidgwick in his *Electronic Theory of Valency* disapproved of the terms 'polar' and 'non-polar' as applied to liquids.[40] In Sidgwick's view these terms were to be applied properly to the description of chemical bonds, and their application to liquids was an unfortunate confusion introduced by G.N. Lewis:[41] the proper classification of liquids was as 'associated' and 'non-associated' liquids.

In 1965 Reichardt[42] showed up the complexity of the situation when he wrote that the polarity of a solvent ... " ... is determined by its solvation behaviour, which in turn depends on the action of intermolecular forces (Coulomb, directional, inductive, dispersion, and charge-transfer forces, as well as hydrogen-bonding forces) between the solvent and the solute". It thus seems unlikely that solvent polarity is potentially capable of being expressed on a straightforward numerical scale, which will be found to be applicable under all circumstances. This is probably why so many different scales of solvent polarity have emerged during the last 20 years, each no doubt with some virtues but also some limitations, and each measuring a blend of different kinds of solvent—solute interaction, characteristic of the particular reference process.

If there are indeed various aspects of solvent polarity, some attempt should be made to isolate and express these individually. This idea is not of recent origin. In connection with the Grunwald—Winstein Y we saw in Section 5.2.1. how the attempt has been made to distinguish ionizing power and nucleophilic participation as factors promoting solvolyses. However, during the last 10 years there have been a variety of attempts to quantify different aspects of solvent polarity and then use the resultant parameters to interpret solvent effects on reactivity through multiple regression. This approach is particularly associated with Koppel and Palm,[23] but others have also made and are making contributions in this field. As long ago as 1968 [43] Chapman, Shorter and colleagues began to use multiple regression on solvent parameters for interpreting solvent effects in the reaction of diazodiphenylmethane and carboxylic acids. In 1971 Fowler, Katritzky, and Rutherford[44] published an extensive survey of the use of multiple regression in connection with solvent effects on spectroscopic, rate, and equilibrium data. More recently Fawcett and Krygowski,[45] and Kamlet and Taft[46] have made extensive contributions. We shall outline some of this work, beginning with the Koppel—Palm treatment.[23]

5.4. THE KOPPEL—PALM TREATMENT

5.4.1. INTRODUCTION AND APPLICATIONS

Solvent—solute interactions are classified as *non-specific* or *specific*. The former are then subdivided into polarization[†] and polarizability effects, to be characterized by the parameters Y and P respectively.[‡] Specific interactions concern donor—acceptor interaction of solvent with solute. A solvent may function as a Lewis base (electron donor) capable of nucleophilic solvation, or as a Lewis acid (electron acceptor) capable of electrophilic solvation. These are to be characterized by the parameters B and E respectively. Scales for these various parameters are devised and may then be used in the correlation analysis of solvent effects by means of equation (5.16):

$$A = A_o + yY + pP + bB + eE \qquad (5.16)$$

A is the value of the solvent-dependent property (log k, ν, *etc.*) in a given solvent, A_o is the statistical quantity (intercept term) corresponding to the value of the property in the gas-phase as reference 'solvent'; y, p, b, and e are the regression coefficients as determined by a least squares procedure. Such an equation can be properly applied only to data for a large number of well-chosen solvents and its success must be examined by proper statistical methods (see Section 2.2). This often limits its application at present.

The setting-up of scales for Y and P depends on the assumption that capability of solvents for polarization and polarizability interactions is validly measured by dielectric constant and refractive index respectively, *i.e.* it must be assumed that these properties, as measured on the bulk solvent, give at least a rough indication of what obtains on the molecular scale (see Section 5.1.3). Thus Y is defined either as $(\varepsilon - 1)/(2\varepsilon + 1)$, the Kirkwood function, or as $(\varepsilon - 1)/(\varepsilon + 1)$, a function based on the expression for molar

‡ Do not confuse with the Grunwald-Winstein, Y.

Footnote for page 146

† Koppel and Palm actually use the word 'polarity' rather than 'polarization'; but in view of the more general meaning of the term polarity, the word polarization seems more appropriate here.

polarization. Except for low values of ε these functions vary approximately collinearly, and the choice of one function rather than another is largely arbitrary. P is appropriately measured by the corresponding functions of the refractive index for sodium light $(n_D^2 - 1)/(2n_D^2 + 1)$ or $(n_D^2 - 1)/(n_D^2 + 2)$. (In the effective numerical range these functions are approximately collinear.)

The choice of scales for B and E presents greater difficulties. We require solvent-sensitive reference processes, which we can regard as measures solely of Lewis base or Lewis acid capability, with non-specific effects playing no part. It seems likely that we can never be absolutely sure about this, but certain processes are known for which we are dealing with just one type of specific effect, and non-specific interactions are effectively nil or are small and can be corrected for.

We consider B first. The OD-stretching frequency of CH_3OD, ν_{OD}, varies greatly with the solvent. This was studied in the late 1930s by Gordy and Stanford,[47,48] more recently by Kagiya *et al.*,[49] and very recently by Burden, Collier, and Shorter.[50] ν_{OD} is believed to be a fairly 'pure' measure of Lewis basicity, and Koppel and Palm base the B scale on the solvent-induced shift from ν_{OD} of gaseous CH_3OD. It is not possible to measure ν_{OD} for CH_3OD dissolved in hydroxylic solvents, so alcohols, *etc.* cannot be directly placed on the B scale. There is a way of dealing with this difficulty to which we shall refer later, after indicating the usefulness of B in the understanding of solvent effects in aprotic solvents.

The definition of a suitable scale of Lewis acidity E presents greater problems. For this purpose we turn to visible spectra. Solvatochromic effects in such spectra are usually complex. It is likely that the processes involved in the Dimroth-Reichardt E_T value[35,36] are simpler than most (Section 5.2.4). In the betaine reference substance the positive centre is well buried in the molecule, whereas the negative centre is exposed. It thus seems probable that the main effect of solvent on the excitation energy is exerted through Lewis acidity interaction with O^- in the ground state. There are some indications, however, that non-specific interactions also influence E_T. To obtain the E scale Koppel and Palm correct the E_T values for these, and adjust the origin of the scale to $E = 0$ for the gas phase. (There is further discussion of the generation of the E scale later.)

Values of ε and n_D are readily available for many solvents; fewer values of B and E are available, and the number of solvents for which both B and E are available is somewhat restricted, but the situation is improving. Values of ε, n_D, B, and E for selected solvents are in Table 5.2. The results of analysing the solvent dependence of a few well-known processes occurring in protic or aprotic solvents are in Table 5.3 and will be discussed briefly. It should be emphasized that full discussion requires detailed consideration of step-wise regression, the values of the standard errors of the regression coefficients, *etc.* (see Section 2.2).

The rates of ionization of t-butyl chloride in a variety of one-component solvents (Table 5.3, 1) are well correlated by Y, P, and E terms. The nucleophilic behaviour of the solvents (the B term) is unimportant. The contribution of eE to the change in log k over the range from gas phase to water is easily the most important. This presumably indicates facilitation of the withdrawal of Cl^- by hydrogen bonding, *etc.* (*cf.* the interpretation of the Grunwald—Winstein Y as a measure of ionizing power; Section 5.2.1.)

The reaction between methyl iodide and tri-n-propylamine in aprotic solvents (Table 5.3, 2) is also well correlated by Y, P, and E terms, with B being unimportant, but here the non-specific effects are dominant. The inclusion of data for alcohols in the regression is unsatisfactory. The alcohols deviate markedly, probably due to their stabilizing the amine reactant by hydrogen bonding.

Log k values for the pyridine-catalysed decomposition of t-butyl peroxyformate in aprotic solvents (Table 5.3, 3) require all four terms, with the dielectric term being dominant. Contributions from E and P terms are also important, but that of B, while at an acceptable level of significance, plays a relatively minor role. This reaction probably proceeds *via* an activated complex (II). It seems that the basic centres in the solvent (*continued on page 152*)

TABLE 5.2. Selected values of Koppel—Palm parameters[a]

Solvent	ε [b]	n_D [b]	E [d]	B
(Gas phase	1.000	1.0000	0	0)
Cyclohexane	2.02	1.4262	(0)	25
Dioxan	2.21	1.4224	4.2	128
Carbon tetrachloride	2.23	1.4603	(0)	31
Benzene	2.28	1.5011	2.1	52
Di-n-butyl ether	3.1	1.3993	(0)	129
Anisole	4.3	1.5170	1.4	75
Chloroform	4.81	1.4180	3.28	39
Chlorobenzene	5.7	1.5218	(0)	45
Tetrahydrofuran	7.4 [c]	1.4076	(0)	145
Dichloromethane	8.9	1.4246	2.7	43
Pyridine	12.3 [c]	1.5100	(0)	260
t-Butyl alcohol	12.5 [c]	1.3848	5.2	125 [e]
i-Propyl alcohol	18.3 [c]	1.3773	8.7	122 [e]
Acetone	20.7 [c]	1.3588	2.1	123
Ethanol	24.3 [c]	1.3614	11.6	117 [e]
Methanol	33.6	1.3286	14.9	114 [e]
Nitrobenzene	34.8 [c]	1.5546	(0)	53
Acetonitrile	37.5	1.3416	5.2	103
NN-Dimethylformamide	38.3	1.4272	2.6	166
Dimethyl sulphoxide	46.6 [c]	1.4783	3.2	192
Water	80.4	1.3330	21.8	90 [e]

a Assembled from refs. 23, 50-52, and 54.

b At 20°C, except as indicated.

c At 25°C.

d (0) means E presumed zero, on structural grounds.

e From correlation of *B* with *B*'; see Section 5.4.2.

TABLE 5.3. Detailed analysis of solvent effects after Koppel and Palm[23]

Reaction; Number of solvents		A_o	$y^{\dagger}$	$p^{\dagger}$	$e^{\dagger}$	$b^{\dagger}$	R	s
1.	33	-12.85	10.21 (±0.98)	13.44 (±3.12	0.275 (±0.015)	-	0.984	0.406
2.	33	-12.83	9.95 (±0.84)	33.1 (±4.3)	0.460 (±0.065)	-	0.946	0.453
3.	17	- 6.76	4.16 (±0.16)	16.03 (±1.06)	0.157 (±0.0105)	0.0025 (±0.0005)	0.995	0.063
4.	11	- 4.14	-	-	-	0.0091 (±0.0012)	0.929	0.256
5.‡	44	- 3.13	4.58 (±0.37)	11.96 (±1.60)	0.195 (±0.020)	-0.0181 (±0.0008)	0.976	0.188
6.	15	- 3.34	4.04 (±1.13)	-	0.207 (±0.071)	0.0145 (±0.0036)	0.932	0.389

1. Log k (sec^{-1}), solvolysis of Bu^tCl, 120°C, in 17 protic and 16 aprotic solvents.

2. Log k ($l\ mol^{-1}\ min^{-1}$), reaction of MeI with Pr^n_3N, 20°C, in 33 aprotic solvents.

3. Log k ($l\ mol^{-1}\ sec^{-1}$), pyridine-catalysed decomposition of t-butyl peroxyformate, 90°C, in 17 aprotic solvents.

4. Log k (sec^{-1}), decomposition of malonic acid, 20°C, in 7 aprotic and 4 protic (mainly amine) solvents.

5.‡ Log k ($l\ mol^{-1}\ min^{-1}$), reaction of benzoic acid and diazodiphenylmethane, 37°C, in 44 aprotic solvents.

6. Log k ($l\ mol^{-1}\ sec^{-1}$), reaction of benzoyl chloride with aniline, 25°C, in 15 aprotic solvents.

The reactions treated in this Table were all featured by Koppel and Palm,[23] but the multiple regressions have been reworked for this book. In some cases it has been possible to include additional data or employ revised values of some solvent parameters. Sources of rate data are given by Koppel and Palm.[23] See also the compilations of rate constants published by Koppel, Palm, and their colleagues at the University of Tartu.

† The regression coefficients of equation (5.16), with $Y = (\varepsilon - 1)/(2\varepsilon + 1)$ and $P = (n_D^2 - 1)/(2n_D^2 + 1)$; standard errors in parentheses.

‡ Ref. 52.

are not able to compete effectively with pyridine in assisting the withdrawal of the proton.

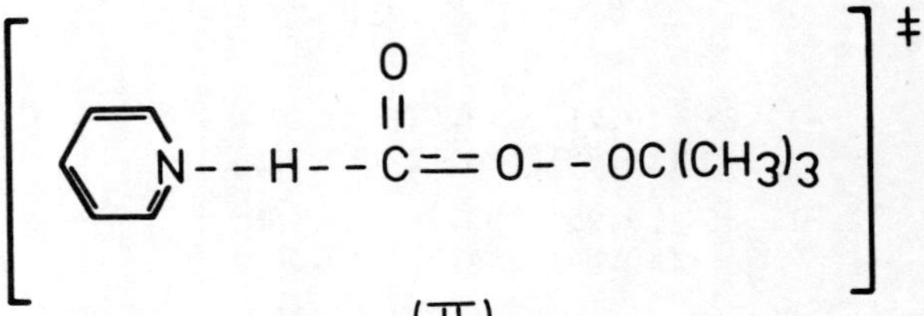

(II)

By contrast, nucleophilic stabilization of the dipolar activated complex (III) is indicated for the decomposition of malonic acid (Table 5.3, 4). Basic solvents may presumably form helpful hydrogen

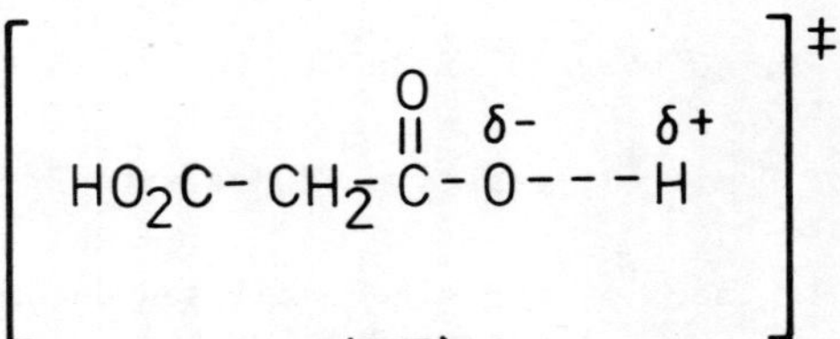

(III)

bonds with the nascent hydrogen ion. Simple regression on *B* accounts for 86% of the variation of log *k*. Although a role for the physical effects corresponding to *Y*, *P*, and *E* may easily be imagined, the inclusion of one or more of these parameters does not significantly improve the regression.

Certain systems show clearly the simultaneous influence of electrophilic and nucleophilic solvation acting either in opposition or to reinforce each other. The reaction between benzoic acid and diazodiphenylmethane (DDM) in aprotic solvents gives significant correlation with all four parameters (Table 5.3, 5). *b* is negative and *e* is positive. This probably indicates nucleophilic stabilization of the carboxylic acid (IV) and electrophilic stabilization of the activated complex (V).

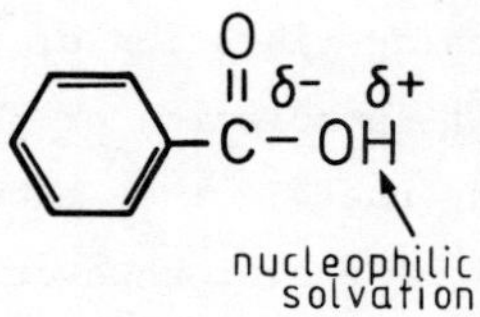

(IV)

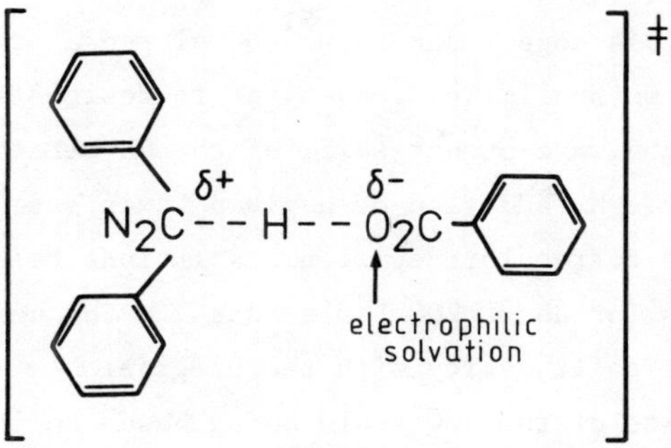

(V)

The reaction of benzoyl chloride with aniline appears to be assisted by both types of solvation (Table 5.3, 6). Possibly the activated complex (VI) requires nucleophilic assistance for the

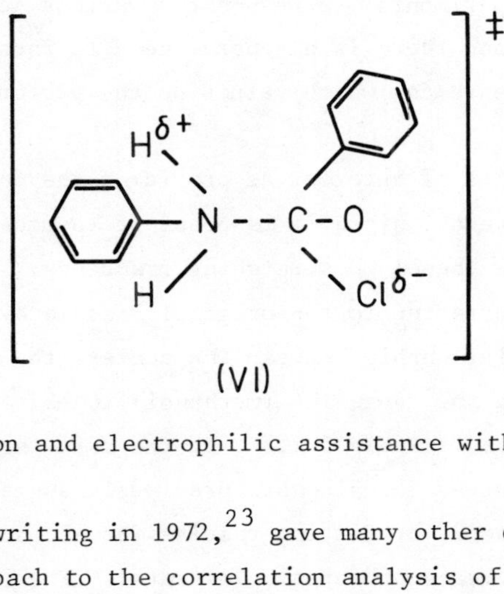

(VI)

removal of the proton and electrophilic assistance with the chloride ion.

Koppel and Palm, writing in 1972,[23] gave many other examples, and their detailed approach to the correlation analysis of solvent effects clearly offered great hope for the future in an area of physical organic chemistry where detailed understanding had been slight. Certainly in the last decade there has been increasing use of multiparameter equations for solvent effects, and a variety of procedures has been adopted. We give first some additional comments on the Koppel—Palm treatment which now seem relevant.

5.4.2. COMMENTS ON THE KOPPEL—PALM TREATMENT

In 1974 Koppel and Paju[51] developed a new solvent basicity scale, which we shall find convenient to designate as the B' scale. This used the wave-number shift of the OH-stretching vibration of phenol, produced by addition of a given Lewis base to a solution of phenol in carbon tetrachloride, dilute solutions being used. They tabulated values for about 200 Lewis bases. The new B' scale, however, shows strong collinearity with the original B (r = 0.987 for 54 solvents), in spite of the new scale being based on interactions in dilute solutions and the original scale being based on interactions in bulk solvent. There are several other possible spectroscopic measures of solvent basicity based on infrared frequencies and usually any one scale is approximately collinear with any other. At present it is difficult to argue for the superiority of any particular scale. Many of the experimental values on the various scales require re-investigation and there is a general need to increase the range and number of solvents for which values on the various scales have been measured.

The B' scale is of interest as providing the means for including hydroxylic solvents, since it is possible to measure the effect of alcohols on the phenol OH-stretching frequency. When the values for alcohols are corrected to the original B scale by using the established relationship between the scales, the B values for alcohols lie in the range 114 (methanol) to 127 (t-pentyl alcohol).[52] (Some values are in Table 5.2.) In the main they are rather lower than the few values for alcohols previously suggested by Koppel and Palm[23] on the basis of fitting results for alcohols to correlations established by means of aprotic solvents. The utility of such values and indeed the validity of combining aprotic and hydroxylic solvents in the same correlations has yet to be properly established (see discussion by Aslam, Collier, and Shorter[52]).

The Koppel—Palm procedure for the parametrization of solvent electrophilicity has come under some criticism, partly because it has been found that the use of E_T directly as a parameter in the multiple regression treatment of solvent effects is often quite successful.

Thus multiple regression on B and E_T of the results for the reaction between DDM and benzoic acid in 44 aprotic solvents gives equation (5.17):

$$\log k = -3.379 - \underset{(\pm 0.00105)}{0.02052\,B} + \underset{(\pm 0.0125)}{0.1293\,E_T} \qquad (5.17)$$

$$R = 0.951 \quad \text{and} \quad s = 0.260$$

Although R is somewhat smaller and s somewhat greater than for the equation based on the Koppel— Palm treatment (Table 5.3, 5) it must be regarded as rather meritorious since it involves only two solvent parameters rather than four. Further it is much more successful than regression on B and $(\varepsilon - 1)/(2\varepsilon + 1)$, which gives $R = 0.907$. E_T is highly correlated with $(\varepsilon - 1)/(2\varepsilon + 1)$ ($r \approx 0.9$), so that after E_T is taken into the correaltion there is no role for $(\varepsilon - 1)/(2\varepsilon + 1)$. However, E_T shows only a low degree of collinearity with $(n_D^2 - 1)/(2n_D^2 + 1)$ ($r \approx 0.29$), so a possible contribution from a polarizability term is worth exploring. Equation (5.18) is found, with $R = 0.969$ and $s = 0.211$ [*cf.* $R = 0.940$ for regression on B, $(\varepsilon - 1)/(2\varepsilon + 1)$, and E].[52]

$$\log k = -5.489 - \underset{(\pm 0.00093)}{0.01877\,B} + \underset{(\pm 0.0102)}{0.1357\,E_T} + \underset{(\pm 1.69)}{8.03}\,\frac{n_D^2 - 1}{2n_D^2 + 1} \qquad (5.18)$$

It thus appears that for the influence of the solvent on the reactions between DDM and benzoic acid it is simpler and reasonably successful to use E_T directly, rather than E and $f(\varepsilon)$. This calls in question the value and reality of Koppel and Palm's analysis of E_T into 'pure' electrophilicity effects and non-specific effects, and also calls in question once again the general validity of applying 'bulk' dielectric constant to molecular phenomena.[53]

We remarked above that E_T is highly correlated with $(\varepsilon - 1)/(2\varepsilon + 1)$ ($r \approx 0.9$) for the solvents involved in the DDM work, and the same is true for any extensive and varied selection of solvents. The solvents in the DDM work include many which on structural grounds would be expected to show appreciable electrophilicity and which do indeed receive significant E values in Koppel and Palm's analysis.

When attention is focused on solvents which would, on structural grounds, not be expected to show appreciable electrophilicity, the relationship between E_T and f(ε) becomes even clearer. This is, in fact, essentially the basis of Koppel and Palm's derivation of E values. In the latest form of the derivation (Koppel and Paju[54]) the E_T values of 26 solvents of assumed zero electrophilicity are correlated with $Y = (\varepsilon - 1)/(\varepsilon + 2)$ and $P = (n_D^2 - 1)/(n_D^2 + 2)$ to give equation (5.19), with $R = 0.974$ and $s = 0.78$.

$$E_T = \underset{(\pm 1.06)}{25.10} + \underset{(\pm 0.74)}{14.84\,Y} + \underset{(\pm 3.70)}{9.59\,P} \tag{5.19}$$

The principal relationship embodied in this equation is between E_T and Y. E values for other solvents are then calculated from equation (5.20).

$$E = E_T - 25.10 - 14.84\,Y - 9.59\,P \tag{5.20}$$

It is difficult not to concede that Koppel and Palm were right *in principle* in seeking to remove from E_T the contributions of non-specific solvent effects as expressed through Y and P. However, their procedure for doing this obviously has its limitations. The regression coefficients and intercept term in equation (5.20) have quite appreciable standard errors, which in turn means that the calculation of an E value involves subtracting from E_T a somewhat uncertain number. For aprotic solvents this number is never much less than E_T itself, the values of E being mainly in the range 0—5.2, so the calculation of E values is subject to considerable uncertainty. Further it must be admitted that the E values of aprotic solvents are not always easily interpretable in terms of molecular structure (*cf.* the B values[50]).

If Koppel and Palm are right in principle, even if the practice is deficient, it then follows that a correlation involving E_T directly must be to some extent artificial, *i.e.* it mixes up non-specific polarization and polarizability effects on the process in question with those on the solvatochromic process used to define E_T. If one wishes to regard correlation with E_T as *fundamental*, then it is necessary also to regard as invalid the use of bulk dielectric constant in the correlation analysis of solvent effects. This is necessary on practical as well as on logical grounds because of the high degree of collinearity between E_T and functions of dielectric constant (see above).

It is widely accepted that E_T is largely governed by solvent Lewis acidity. This interpretation of E_T seems particularly appropriate for protic solvents, for which hydrogen bonding to the betaine O^- may be envisaged, and the values all lie at the top of the E_T scale, *viz* 44—63. The situation for the aprotic solvents, whose E_T values lie in the range 31—46 is much less clear and it may well be that E_T measures various aspects of intermolecular interactions, *e.g.* dipole—dipole, donor—acceptor, dispersion forces *etc*. Viewed in this way it is perhaps not unreasonable to regard E_T as a kind of substitute for dielectric and (perhaps) refractive index parameters, remedying to some extent the deficiencies of these 'bulk' properties when applied to phenomena on the molecular scale.[53]

When the results of the Koppel—Palm analysis of the reaction of benzoic acid with DDM in aprotic solvents are examined in detail, it appears that the analysis places certain pairs of solvents in the 'wrong' order, *e.g.* the experimental values of k are in the order $CCl_4 \ll C_6H_6$, while the regression gives $CCl_4 \gg C_6H_6$. No doubt such relationships are due to special effects of which the parameters used in the correlation take no account.† The question is sometimes raised as to whether it is possible to parametrize additional special effects.[53]

Two possible factors come readily to mind. First, there is the possible occurrence of steric effects on nucleophilic solvation which are not included in B because B is based on interaction with the rather small molecule CH_3OD. Secondly, solvents vary quite considerably in viscosity and this could conceivably influence the rate at which the reactants come together. One may reasonably

† Indeed in the ultimate analysis each solvent is just itself. True though this remark undoubtedly is, it is not particularly helpful to the process of correlation analysis!!

parametrize the former by molar refraction as a measure of molecular volume, and the latter by the logarithm of the fluidity. The trouble is that these quantities turn out to be considerably collinear with some of the normal Koppel— Palm parameters, *e.g.* molar refraction is considerably collinear with both E and $(n_D^2 - 1)/(2n_D^2 + 1)$. If, in stepwise regression, E and $(n_D^2 - 1)/(2n_D^2 + 1)$ are preferred (albeit marginally), no significant role is left for molar refraction. The same remark applies to log (fluidity), and may well apply to any other effects that one might consider in a supplementary way. Thus the Koppel— Palm analysis is, so to say, the end of the road in that particular direction. There is nothing further to be done in explaining residual variation of the dependent variable. If we wish to have correlation expressions which are more successful, then these must be based on some other model, maybe totally different or maybe parametrizing the important solvent properties of Lewis basicity and acidity in a different way.

5.4.3. ALTERNATIVE SOLVENT PARAMETERS

We have already mentioned the existence of several alternative scales of solvent basicity based on infrared frequencies and the possible direct use of E_T as a practical measure of electrophilicity and certain other intermolecular interactions.

In addition to the ir quantities, various other solvent properties have been suggested as measures of basicity. For example there are the parameters related to the interaction of *p*-fluorophenol and bases, developed by Taft, Arnett and their co-workers.[55-57] These are the formation constants of the hydrogen-bonded complexes in carbon tetrachloride (used as log K_f) and their enthalpies of formation, ΔH_f. In an attempt to apply these quantities to the results for DDM reacting with benzoic acid, the author and his colleagues found that values of log K_f were available for only 14 of the solvents and of ΔH_f for only 18 of them.[53]

Koppel— Palm type correlations in which log K_f or ΔH_f replaced B were done for the numbers of solvents mentioned, with parallel ordinary Koppel— Palm correlations for the same selections of solvents for comparison. The conclusion was that for the solvents in question log K_f or ΔH_f did not reveal any great advantage over B. Of course the comparison would have been more satisfactory if it could have been done for the full range of 44 solvents.

Gutmann[58] has developed other parameters for solvent nucleophilicity and electrophilicity known as the donor number (DN) and acceptor number (AN) respectively. The DN is defined as $-\Delta H/\text{kcal mol}^{-1}$ for the formation of a complex between the donor and antimony(V) chloride in 1,2-dichloroethane. The AN is defined in terms of the ^{31}P nmr shifts produced in triethylphosphine oxide by electrophilic solvent actions, which lower the electron density at P by coordination with the oxygen atom. The shifts are related to that for the $Et_3PO \rightarrow SbCl_5$ adduct in 1,2-dichloroethane, which is arbitrarily assigned a value of 100. Values of DN are known for *ca.* 50 solvents and of AN for *ca.* 30, but when the question of applying them in the DDM work was considered, it was found that values of both DN *and* AN were known for only 10 of the solvents involved.[53]

For these solvents stepwise regression of log k on DN, AN, and $(\varepsilon - 1)/(2\varepsilon + 1)$ was performed, and a normal Koppel—Palm regression was done for comparison. DN was taken first ($r = 0.806$), then AN ($R = 0.970$) and finally $(\varepsilon - 1)/(2\varepsilon + 1)$, giving $R = 0.987$ for the final three-parameter equation. The preference for AN rather than $(\varepsilon - 1)/(2\varepsilon + 1)$ ($R = 0.940$) was only marginal, and possibly an artefact of the particular limited number of solvents involved. In the normal stepwise regression for these solvents the terms were taken in the usual order B, $(\varepsilon - 1)/(2\varepsilon + 1)$, E and $(n_D^2 - 1)/(2n_D^2 + 1)$ and the corresponding correlation coefficients are 0.867, 0.948, 0.975, and 0.993. Thus B by itself is more successful than DN alone but the addition of AN effects an improvement considerably greater than that produced by $(\varepsilon - 1)/(2\varepsilon + 1)$ in the Koppel—Palm type of regression. Possibly AN will be very valuable as a parameter of solvent electrophilicity when values are available for many more solvents.

Fawcett and Krygowski[45] suggested the correlation analysis of solvent effects in terms of DN and E_T. They used early and therefore restricted results on the DDM reaction in such a multiple regression (8 solvents) giving $R = 0.989$. This was stated to be a better correlation than the Koppel—Palm correlation of our results at that time, which gave $R = 0.975$. However, this unfavourable comparison ignored the much greater number of data points involved in establishing our equation of that time, namely 24.

In the next Section we shall discuss the new parametrization of solvent properties proposed by Kamlet and Taft.

5.4.4. THE KAMLET—TAFT TREATMENT

In a sense this is developing as a rival to the Koppel—Palm treatment, although the fundamental ideas of the two treatments have much in common, *i.e.* that it is necessary to consider specific and non-specific solvent—solute interactions, and that the former should be subclassified as solvent Lewis basicity and Lewis acidity interactions. The Kamlet—Taft treatment has been developed through a long series of papers since 1975; a detailed summary and discussion of the present position, in the view of the principal investigators, has been given recently.[46] Here the treatment can only be outlined.

The authors refer to their approach as the *solvatochromic comparison method*. They develop a β scale of solvent hydrogen-bond acceptor (HBA) basicities (corresponding to the Koppel—Palm B scale) and an α scale of solvent hydrogen-bond donor (HBD) acidities (corresponding to the Koppel—Palm E scale) on the basis of uv/visible spectral data. They also develop a π^* scale of 'solvent polarity' for solvent solute interactions of the non-specific kind (note the rather special use of the term 'solvent polarity'). This corresponds to the use of dielectric constant and refractive index in the Koppel—Palm treatment as measures of polarization and polarizability interactions. Functions of ε and n_D^2 play no part in the Kamlet—Taft treatment. The standard substrates involved in establishing the scales of β, α, and π^* are mainly nitrophenols and nitroanilines, and their OR or NR_2 derivatives, assumptions being made concerning the various solvent—solute interactions which are possible with the various substrates. By solvatochromic comparison of suitably related substrates, particular types of interaction can be isolated; (*cf.* Taft's analysis of polar, steric, and resonance effects in ester hydrolysis, Chapter 4).

The setting-up of the β scale is based on determining enhanced solvatochromic shifts in HBA solvents for 4-nitroaniline relative to *NN*-diethyl-4-nitroaniline. The interactions of these two substrates with the solvents are considered to differ qualitatively only in the possibility of hydrogen bond formation involving the amino-hydrogens of the former. The enhanced shift is therefore a measure of the HBA capability of the solvent. A similar comparison of 4-nitrophenol and 4-nitroanisole is possible. A number of such comparisons is considered and the β scale emerges from an averaging process involving normalization of the enhanced shifts to those for hexamethylphosphoramide (HMPA) as solvent, *i.e.* for this strongly basic solvent, β is taken as 1.000.

The α scale is based on the enhanced solvatochromic shifts in HBD solvents for the *N*-phenol-pyridinium betaine dye [(I), R = H; the compound involved in the Dimroth-Reichardt E_T (30) scale; see 5.2.4.] compared with 4-nitroanisole. The interactions of these two substrates with the solvents are considered to differ significantly only in the great importance of hydrogen bonding between the exposed O^- in the betaine and HBD solvents. The enhanced shift is therefore a measure of the HBD capability of the solvent. Methanol, a strong HBD solvent, is taken as a reference point for which α = 1.000.

The π* scale of 'solvent polarity' (see above) is so called because it is based on the effects of solvents on the $p \rightarrow \pi^*$ and $\pi \rightarrow \pi^*$ electronic spectral transitions of various nitroaromatic compounds, *e.g.* 4-nitroanisole, 1-ethyl-4-nitrobenzene, *NN*-diethyl-4-nitroaniline, *etc.* Seven such 'primary' indicators were employed in the initial construction of the π* scale, which was then expanded and refined by considering solvent effects on about 40 more substrates. A normalized range of 0.000 for cyclohexane to 1.000 for DMSO was chosen.

A selection of β, α, and π* values for common solvents is shown in Table 5.4. It must be emphasized that the significance of these values can only be properly appreciated through a careful study of

TABLE 5.4. Selected values of Kamlet— Taft parameters[a]

Solvent	π^*	β	α
Cyclohexane	0.00	nil	nil
Triethylamine	0.14	0.71	nil
Di-n-butyl ether	0.24	0.46	nil
Carbon tetrachloride	0.29	nil	nil
t-Butyl alcohol	0.41	1.01	0.62
i-Propyl alcohol	0.46	0.95	0.78
Ethanol	0.54	0.77	0.86
Dioxan	0.55	0.37	nil
Ethyl acetate	0.55	0.45	nil
Tetrahydrofuran	0.58	0.55	nil
Benzene	0.59	0.10	nil
Methanol	0.60	0.62	0.98
Acetone	0.68	0.48	0.07
Chlorobenzene	0.71	0.07	nil
Chloroform	0.76	nil	0.34
Acetonitrile	0.76	0.31	0.15
Dichloromethane	0.80	nil	0.22
Pyridine	0.87	0.64	nil
NN-Dimethylformamide	0.88	0.69	nil
Benzonitrile	0.90	0.41	nil
Dimethyl sulphoxide	1.00	0.76	nil

a It must be emphasized that these values are for exemplification only. The authors[46] regard them as of very variable reliability and many are liable to be changed.

Kamlet and Taft's summarizing article[46] and the original papers. These authors have shown that β, α, and π^* may be applied in the correlation analysis by multiple regression of reaction rates and equilibria, and of spectroscopic data; equation (5.21) applies.[‡]

‡ The quantities A, A_o, and p in equation (5.21) and in Table 5.5 are represented by other symbols in Kamlet and Taft's papers, but the above are convenient for the present discussion.

$$A = A_o + p\pi^* + a\alpha + b\beta \qquad (5.21)$$

Some typical examples taken from the summarizing article[46] are in Table 5.5. Several of the reactions also appear in Table 5.3.

TABLE 5.5. Detailed analysis of solvent effects after Kamlet and Taft[46]

Reaction	Number of solvents	A_o	p	a	b	R	s
1	17	5.6	8.54	7.06	-	0.992	0.65
2	44	-4.24	4.43	-	-	0.981	0.21
3	15	0.30	1.96	-	-	0.986	0.11
4	17	-0.78	-	-	38.4	0.982	1.86
5	12	0.13	3.06	-	-5.63	0.989	†
6	8	-5.83	3.11	1.94	-	0.991	0.18

1. Solvolysis of t-BuCl, 120°C, in 11 aliphatic HBD and 6 non-HBD solvents. Dependent variable is 1.800 [log k(solvent) - log k (gas phase)] kcal mol^{-1}, *cf.* reaction 1 in Table 5.3.

2. Log k (1 mol^{-1} min^{-1}), reaction of MeI with Pr^n_3N, 20°C, in 44 non-HBD solvents, *cf.* reaction 2 in Table 5.3.

3. 3+ log k (1 mol^{-1} sec^{-1}), pyridine-catalysed decomposition of t-butyl peroxyformate, 90°C, in 15 non-HBD (or weak-HBD) solvents, *cf.* reaction 3 in Table 5.3.

4. Gutmann's donor number, DN, for 17 HBA solvents.

5. Log k (1 mol^{-1} min^{-1}), reaction of benzoic acid with diazodiphenylmethane, 37°C, in 12 aliphatic non-HBD solvents, *cf.* reaction 5 in Table 5.3.

6. Log k (1 mol^{-1} sec^{-1}), reaction of 2-thiophenesulphonyl chloride with aniline, 25°C, in 8 aliphatic protic and aprotic solvents.

† Not given.

The Kamlet—Taft and Koppel—Palm treatments of solvent effects for the ionization of t-butyl chloride (reaction 1 in both Tables) agree that the Lewis acidity of the solvent is important and that solvent basicity is unimportant. The 'polarity' parameter π* successfully fulfils the role played by dielectric and refractive index functions in the Koppel—Palm treatment.

π^* by itself seems highly successful in accounting for the solvent effect in the Menschutkin reaction (reaction 2 in both Tables), and this situation contrasts with the greater complexity and apparently more modest success of the Koppel—Palm treatment, whether measured by R or s. π^* by itself is also reasonably successful for the pyridine-catalysed decomposition of t-butyl peroxyformate (reaction 3), and the greater success of the Koppel—Palm treatment is only accomplished by involving all four solvent parameters.

The Gutmann donor number DN[58] (based on complex formation with antimony (V) chloride; see Section 5.4.3.) is well correlated by Kamlet—Taft β for the 17 HBA solvents selected (Table 5.5, 4) although other such solvents are said to deviate.

The reaction of 2-thiophenesulphonyl choride with aniline (Table 5.5, 6) shows the importance of Lewis acidity in assisting the departure of the chloride leaving group, as well as the importance of general polarity as measured by π^* (*cf.* reaction 6, in Table 5.3, between benzoyl chloride and aniline).

Kamlet and Taft present an analysis of the reaction between DDM and benzoic acid, limited to 12 aliphatic non-HBD solvents (Table 5.5, 5). They comment that the reaction constitutes one of the few cases in which the π^* and β terms act in opposition. The Koppel—Palm analysis for this reaction may be much more extensive, because the necessary solvent parameters are available (Table 5.3, 5). In order to enable a full comparison of the Kamlet—Taft and Koppel—Palm treatments of this reaction to be made, the present author and his colleagues[59] have recently (re-)determined α, β, and π^* values for all 44 aprotic solvents involved in the work on the reaction between diazodiphenylmethane and benzoic acid (see Sections 5.4.1, 2, and 3, and Table 5.3).[52,53] Multiple regression of log k (37°C) on these solvent parameters gives:[59]

$$\log k = 0.2038 + \underset{(\pm 0.2092)}{2.7139\alpha} - \underset{(\pm 0.1216)}{3.7049\beta} + \underset{(\pm 0.1096)}{1.2134\pi^*} \qquad (5.22)$$

$$n = 44 \qquad R = 0.980 \qquad s = \pm 0.1708$$

This equation may be compared with the relevant equation in Table 5.3, equation (5.17) and other correlation equations for the solvent effect on this reaction. In this case the Kamlet—Taft treatment seems to give results comparable with those of the Koppel—Palm treatment.

No doubt there will be much discussion in the years ahead of the relative merits of the parametrizations used by the various groups of authors. The most important feature of the present situation is the undoubted success of several forms of multiparameter approach to solvent effects by comparison with any approach based on simple linear regression on individual measures of 'solvent polarity'.

A possible development is indicated by recent work of Japanese authors on the rates of the reaction between diazodiphenylmethane and tetracyanoethylene to give 1,1-diphenyl-2,2,3,3-tetracyanocyclopropane.[60] As for the reaction between DDM and benzoic acid, the Lewis basicity of the solvent seems to be the most important factor governing the rate. However, the two reactions do not show a particularly good mutual LFER, and in a series of solvents, orders of solvent basicity which differ in detail appear to operate in the two reactions. It is suggested that this is because benzoic acid is a 'hard' protic acid whereas TCNE is a 'soft' Lewis acid.[61] Thus a comprehensive treatment of solvent effects could involve considering a 'sliding-scale' of solvent basicity, *cf.* the sliding-scale of σ values, Section 3.1.1.

This is an appropriate point to mention that modern sophisticated statistical techniques are being applied to solvent effects. See, for example, papers by Chastrette[62] and by Sjöström and Wold.[63]

5.5. MISCELLANEOUS TOPICS

5.5.1. THE CORRELATION ANALYSIS OF SOLVENT EFFECTS BY MEANS OF SUBSTITUENT CONSTANTS

It is sometimes possible to express aspects of solvent polarity in a series of structurally related solvents by means of appropriate substituent constants, thus providing a link between solvent effects and the Hammett or Taft equations. In a general way this may not be entirely straightforward, since substituent constants concern effects transmitted to a localized reaction centre, while solvent effects, in principle, involve the interaction of the reacting species with the solvent molecule in its entirety. In some series of solvents, however, a particular centre in the solvent molecule is likely to dominate solvent—solute interactions and in such cases the influence of substituents on that centre may play an important role in solvent effects. Electronic or steric substituent constants may then be relevant as solvent parameters, usually in connection with multiple regression.

One of the best established correlations of this kind is for the reaction of DDM with benzoic acid in a series of alcohols. The explanatory variables are σ^* for the alkyl group of the alcohol, the Kirkwood function $f(\varepsilon) = (\varepsilon - 1)/(2\varepsilon + 1)$ and $n_{\gamma H}$, the number of γ-hydrogen atoms in the alcohol. Log k values (30°C) are now available for this reaction in a considerable number of alcohols. At present the correlation equation is appropriately based on the results for 22 alcohols, mainly simple aliphatic primary, secondary, and tertiary alcohols, but including certain phenyl-substituted alcohols such as benzyl alcohol. The equation is:[64]

$$\log k = -1.484 + \underset{(\pm 0.474)}{3.957 f(\varepsilon)} + \underset{(\pm 0.099)}{3.055\sigma^*} + \underset{(\pm 0.0069)}{0.029 n_{\gamma H}} \qquad (5.23)$$

$$n = 22 \qquad R = 0.993 \qquad s = 0.0525$$

The σ^* term is the most important, in that σ^* is the first parameter to enter in stepwise regression, and simple regression of log k on σ^* gives $r = 0.958$. The importance of the σ^* term confirms the role of nucleophilic and electrophilic solvation, which was indicated by the Koppel—Palm treatment of the same reaction in aprotic solvents; see Section 5.4.1. In particular, hydrogen bonding involving RO$\underline{\text{H}}$ may be seen as an important factor stabilizing the nascent benzoate ion in the activated complex (structure V in Section 5.4.1). The $n_{\gamma H}$ term is seen as a measure of a steric effect which moderates the basic properties of the alcohol O. It is responsible for the reactivity in an alcohol such as 2-methylpropan-1-ol ($n_{\gamma H} = 6$) being considerably greater than that in other alcohols of similar σ^* and ε. This effect associated with γ-hydrogen atoms has recently been confirmed by studies involving several highly-branched alcohols.[65] It has also been confirmed that equation (5.23) applies fairly well to the reaction when carried out in various negatively substituted alcohols, notably trifluoroethanol.[65]

Further examples of this type of correlation analysis of solvent effects were given by Koppel and Palm.[23]

5.5.2. SOLVENT EFFECTS ON ρ AND σ VALUES

The dependence of the Hammett reaction constant ρ (Chapter 3) on the solvent is often very marked and has been discussed since the earliest days of LFER. Hammett himself[66] predicted a linear relationship with the reciprocal of the dielectric constant. For some reactions this holds approximately within a limited range of solvents, but extending the range often reveals marked deviations.

Thus for diazodiphenylmethane reacting with *meta*- or *para*-substituted benzoic acids in aliphatic alcohols ρ increases as ε decreases.[43] This may be interpreted in two ways. First, the transmission of the polar effects of the substituents through the surrounding medium increases in importance relative to transmission through the molecular cavity as ε is decreased. Alternatively we may consider the charge separation in the transition state relative to that in the initial state. At high dielectric constant the energy necessary to bring about the charge separation is relatively small and this gives rise to a low susceptibility to the polar effects of substituents. The ρ value thus increases as the energy necessary to achieve the charge separation increases as ε is decreased. This implies an operation of the reactivity—selectivity principle[67] and an approximate inverse relationship between ρ and log k^{o} (parent system, Section 3.1) is in fact found.

However, benzyl alcohol deviates very markedly from the relationship between ρ and ε, and the attempt to include aprotic solvents in the same relationship is largely a failure. It is clear that ρ values, like log k values for given reactants, are influenced both by non-specific and by specific solvent effects. Widespread validity of simple correlations is not to be expected and multiple regression of ρ on suitable solvent parameters is required.

Koppel and Palm[23] discussed the solvent dependence of ρ values in connection with their equation (5.16). We will give an example arising from more recent work. Values of ρ are now available for the reaction of DDM with *meta*-substituted benzoic acids in eleven aprotic solvents at 30°C.[68] The regression equation for a Koppel—Palm type analysis is:

$$\rho \;=\; 3.341 \;-\; \underset{(\pm 0.0011)}{0.0056B} - \underset{(\pm 0.403)}{1.752\,\frac{\varepsilon - 1}{2\varepsilon + 1}} \;-\; \underset{(\pm 0.0245)}{0.0351E} \qquad (5.24)$$

with R = 0.967 and s = 0.108. The order of decreasing significance of the terms is B, f(ε), E, the correlation coefficients for the successive one- and two-parameter equations being 0.873 and 0.957 respectively. The contribution made by the E term is only of marginal significance, as E is unfortunately appreciably collinear ($r \approx 0.5$) with B for the solvents in question. The correlation analysis can also be done with B and the Dimroth-Reichardt E_T parameter as explanatory variables. The regression equation is:

$$\rho \;=\; 4.516 \;-\; \underset{(\pm 0.0008)}{0.0053B} \;-\; \underset{(\pm 0.009)}{0.0494E_T} \qquad (5.25)$$

with R = 0.973 and s = 0.092. (Collinearity of B and E_T is appreciable; $r \approx 0.5$.) Comparison of equation (5.24) and Table 5.3, 5, and of equations (5.25) and (5.17) shows that an increase in Lewis basicity of the solvent decreases both ρ and log k (for reaction of benzoic acid itself), while an increase in dielectric constant and Lewis acidity decreases ρ but increases log k. Opposite influences of a given factor on ρ and log k would be expected from a simple application of the reactivity-selectivity principle, whereas the parallel effects on ρ and log k would not be predicted.

The ρ values for the DDM reactions in eleven alcohols may be subjected to correlation analysis through an equation analogous to equation (5.23), namely

$$\rho \;=\; 2.362 \;-\; 3.138\mathrm{f}(\varepsilon) \;-\; 0.910\sigma^* \;-\; 0.0141n_{\gamma H} \qquad (5.26)$$

with R = 0.993 and s = 0.025. It will be noted that opposite signs are associated with corresponding terms in equations (5.23) and (5.26).

A quantitative analysis of solvent effects on reaction constant has recently been presented by Poh.[69] The systems treated include the dissociation equilibria of several series of organic acids and also the reactions of DDM with various series of acids. The operation of several factors is considered: 'effective' dielectric constant (as in Kirkwood-Westheimer theory[70]), solvent hydrogen-bonding ability, conformation of the acids, and, for the DDM reactions, variation in the charge developed in the transition state. The treatment appears to gain a modest success, particularly for the dissociation equilibria.

Values of other reaction constants such as ρ^* and δ (Chapter 4) and ρ_I and ρ_R (Chapter 3) may also show solvent dependence but there has been little systematic study of such matters. Koppel and Palm[23] gave some examples of the application of multiple regression in these cases.

Variation of σ values with solvent has been detected for certain substituents, as indicated by the qualifying footnotes that sometimes appear in compilations of σ values. There has been little systematic study of this, and many of the data do not permit reaction-dependence and solvent-dependence of σ to be clearly distinguished. Much of the evidence for solvent-dependence of σ is for substituents such as OH and OMe, and it seems likely that hydrogen-bonding interactions between substituent and solvent are important.

Aslam *et al.*[68] have recently obtained σ values for *m*-OH and *m*-OMe by studies of the reaction between DDM and substituted benzoic acids in several solvents. In 7 alcohols σ for *m*-OH ranges from -0.087 to -0.126, but in acetone and in ethyl acetate the values are less negative, -0.047 and -0.035 respectively. These values contrast with the usually quoted σ value of +0.12 for *m*-OH from the dissociation of benzoic acids in water and a value of +0.05 from an ion-pair equilibrium of benzoic acids in benzene.[70] Values of σ for *m*-OMe from the DDM reaction in 7 alcohols, ethyl acetate, and acetone are less solvent-dependent, ranging from +0.021 to +0.048, and may be contrasted with the value of +0.12 from the dissociation of benzoic acids in water. A detailed interpretation is suggested in terms of the modifying of the electronic effect by various forms of solvent—substituent hydrogen bonding interaction.

5.6 REFERENCES

1 M. Berthelot and L. P. de Saint-Gilles, *Ann. Chim. et Phys.*, 3 *Ser.*, *65*, 385 (1862); *66*, 5 (1862); *68*, 225 (1863).

2 N. A. Menschutkin, *Zeit. Phys. Chem.*, *1*, 611 (1887).

3 N. A. Menschutkin, *Zeit. Phys. Chem.*, *6*, 41 (1890).

4 L. Claisen, *Liebigs Ann. Chem.*, *291*, 25 (1896).

5 W. Wislicenus, *Liebigs Ann. Chem.*, *291*, 147 (1896).

6 L. Knorr, *Liebigs Ann. Chem.*, *293*, 70 (1896).

7 A. Hantzsch and O. W. Schultze, *Ber. Deutsch. Chem. Ges.*, *29*, 2251 (1896).

8 T. S. Patterson and H. H. Montgomerie, *Trans. Chem. Soc.*, 26 (1912).

9 C. N. Hinshelwood, *The Kinetics of Chemical Change in Gaseous Systems*, Clarendon Press, Oxford, 1926.

10 E. A. Moelwyn-Hughes, *The Kinetics of Reactions in Solution*, Clarendon Press, Oxford, 1933.

11 Ref. 10, pp.106—119.

12 Summary in S. Glasstone, K. J. Laidler, and H. Eyring, *The Theory of Rate Processes*, McGraw-Hill, New York, 1941, Chapter 8.

13 M. Richardson and F. G. Soper, *J. Chem. Soc.*, 1873 (1929).

14 F. G. Soper and E. Williams, *J. Chem. Soc.*, 2297 (1931).

15 E. D. Hughes and C. K. Ingold, *J. Chem. Soc.*, 244 (1935).

16 C. K. Ingold, *Structure and Mechanism in Organic Chemistry*, Bell, London, 1st edn., 1953; 2nd edn., 1969.

17 Ref. 16, 1st edn., p.347. Quotation by kind permission of the publisher.

18 K. J. Laidler, *Reaction Kinetics: Volume II — Reactions in Solution*, Pergamon, Oxford, 1963.

19 J. G. Kirkwood, *J. Chem. Phys.*, *2*, 351 (1934).

20 E. M. Kosower, *An Introduction to Physical Organic Chemistry*, Wiley, New York, 1968, p.272.

21 Chr. Reichardt, *Solvent Effects in Organic Chemistry*, Verlag Chemie, Weinheim, 1979, Chapter 7.

22 T. R. Griffiths and D. C. Pugh, *Coord. Chem. Rev.*, *29*, 129 (1979).

23 I. A. Koppel and V. A. Palm in *Advances in Linear Free Energy Relationships*, N. B. Chapman and J. Shorter, editors, Plenum, London, 1972, Chapter 5.

24 E. Grunwald and S. Winstein, *J. Amer. Chem. Soc.*, *70*, 846 (1948).

25 Ref. 21, pp.231–5.

26 S. Winstein, E. Grunwald, and H. W. Jones, *J. Amer. Chem. Soc.*, *73*, 2700 (1951).

27 T. W. Bentley, F. L. Schadt, and P. v. R. Schleyer, *J. Amer. Chem. Soc.*, *94*, 992 (1972).

28 F. L. Schadt, T. W. Bentley, and P. v. R. Schleyer, *J. Amer. Chem. Soc.*, *98*, 7667 (1976).

29 C. G. Swain, R. B. Moseley, and D. E. Bown, *J. Amer. Chem. Soc.*, *77*, 3731 (1955).

30 For some recent developments involving Y *etc*. see: T. W. Bentley, C. T. Bowen, W. Parker, and C.I.F. Watt, *J. Amer. Chem. Soc.*, *101*, 2486 (1979); T. W. Bentley, C. T. Bowen, H. C. Brown, and F. J. Chloupek, *J. Org. Chem.*, *46*, 38 (1981).

31 M. Gielen and J. Nasielski, *J. Organometal. Chem.*, *1*, 173 (1963); *7*, 273 (1967).

32 C. Lassau and J.-C. Jungers, *Bull. Soc. Chim. France*, 2678 (1968).

33 E. M. Kosower, *J. Amer. Chem. Soc.*, *80*, 3253 (1958).

34 Ref. 20, p.296.

35 K. Dimroth, Chr. Reichardt, T. Siepmann, and F. Bohlmann, *Ann. Chem.*, *661*, 1 (1963).

36 Chr. Reichardt, *Angew. Chemie Int. Edn. in Engl.*, *18*, 98 (1979).

37 S. Brownstein, *Canad. J. Chem.*, *38*, 1590 (1960).

38 L. Carroll, *Alice's Adventures in Wonderland and Through the Looking-Glass*, edited with an introduction by R. L. Green, O.U.P., London, 1971, (*a*) p.136, (*b*) p.190.

39 V. Gold, *Pure Appl. Chem.*, *51*, 1725 (1979).

40 N. V. Sidgwick, *The Electronic Theory of Valency*, O.U.P., London, 1927, p.133.

41 G. N. Lewis, *J. Amer. Chem. Soc.*, *35*, 1448 (1913).

42 Chr. Reichardt, *Angew. Chem. Int. Edn. in Engl.*, *4*, 29 (1965).

43 A. Buckley, N. B. Chapman, M.R.J. Dack, J. Shorter, and H. M. Wall, *J. Chem. Soc. (B)*, 631 (1968).

44 F. W. Fowler, A. R. Katritzky, and R.J.D. Rutherford, *J. Chem. Soc.*, *(B)*, 460 (1971).

45 W. R. Fawcett and T. M. Krygowski, *Austral. J. Chem.*, *28*, 2115 (1975).

46 A summarizing review has appeared in M. J. Kamlet, J.L.M. Abboud, and R. W. Taft, *Progr. Phys. Org. Chem.*, *13*, 485 (1981).

47 W. Gordy and S. C. Stanford, *J. Chem. Phys.*, *9*, 204 (1941).

48 W. Gordy, *J. Chem. Phys.*, *9*, 215, 440 (1941).

49 T. Kagiya, Y. Sumida, and T. Inoue, *Bull. Chem. Soc. Japan*, *41*, 767 (1968).

50 A. G. Burden, G. Collier, and J. Shorter, *J. Chem. Soc. Perkin Trans. II*, 1627 (1976).

51 I. A. Koppel and A. I. Paju, *Org. React.*, *11*, 121 (1974).

52 M. H. Aslam, G. Collier, and J. Shorter, *J. Chem. Soc. Perkin Trans. II*, 1572 (1981).

53 A. G. Burden, N. B. Chapman, H. F. Duggua, and J. Shorter, *J. Chem. Soc. Perkin Trans. II*, 296 (1978).

54 I. A. Koppel and A. I. Paju, *Org. React.*, *11*, 137 (1974).

55 D. Gurka and R. W. Taft, *J. Amer. Chem. Soc.*, *91*, 4794 (1969).

56 E. M. Arnett, L. Joris, E. Mitchell, T.S.S.R. Murty, T. M. Gorrie, and P. v. R. Schleyer, *J. Amer. Chem. Soc.*, *92*, 2365 (1970).

57 E. M. Arnett, E. J. Mitchell, and T.S.S.R. Murty, *J. Amer. Chem. Soc.*, *96*, 3875 (1974).

58 V. Gutmann, *Coord. Chem. Rev.*, *18*, 225 (1976) and references therein.

59 G. Collier, I. A. Pickering, and J. Shorter, to be published.

60 T. Oshima, S. Arikata, and T. Nagai, *J. Chem. Res.*, *(S)* 204, (*M*) 2518 (1981).

61 R. G. Pearson, *J. Chem. Ed.*, *45*, 581, 643 (1968). (See also Chapter 6.)

62 M. Chastrette, *Tetrahedron*, *35*, 1441 (1979).

63 M. Sjöström and S. Wold, *J. Chem. Soc. Perkin Trans. II*, 104 (1981).

64 M. H. Aslam, A. G. Burden, N. B. Chapman, J. Shorter, and M. Charton, *J. Chem. Soc. Perkin Trans. II*, 500 (1981).

65 D. Mather and J. Shorter, to be published.

66 L. P. Hammett, *J. Amer. Chem. Soc.*, *59*, 96 (1937).

67 H. C. Brown and K. L. Nelson, *J. Amer. Chem. Soc.*, *75*, 6292 (1953).

68 M. H. Aslam, N. B. Chapman, M. Charton, and J. Shorter, *J. Chem. Res.*, *(S)* 188, (*M*) 2301 (1981).

69 B.-L. Poh, *Austral. J. Chem.*, *33*, 1175 (1980).

70 J. G. Kirkwood and F. H. Westheimer, *J. Chem. Phys.*, *6*, 506, 513 (1938).

71 M. M. Davis and H. B. Hetzer, *J. Res. Nat. Bur. Stand.*, *60*, 569 (1958).

CHAPTER 6

The Role of the Reagent

	Page
6.1. Introduction	173
6.2. Acid-base catalysis: the Brönsted equation	174
6.2.1. Some forms of acid catalysis	174
6.2.2. The Brönsted equation	176
6.3. Nucleophilicity in bimolecular reactions	182
6.3.1. Brönsted-type relationships	182
6.3.2. The Swain—Scott equation	186
6.3.3. The Edwards equation	191
6.3.4. Hard and soft acids and bases	193
6.3.5. Ritchie's work	195
6.3.6. Jencks's equation	199
6.3.7. Miscellaneous matters and conclusion	200
6.4. References	204

6.1. INTRODUCTION

In Chapters 3 and 4 we dealt with correlation analysis in relation to the *substrate* of an organic reaction, *i.e.* we discussed the influence of substituents on the reactivity of a given functional group undergoing reaction with a given reagent. It is also possible to apply correlation analysis to reaction series involving a given substrate reacting with various reagents. One aspect of this has already been examined in Chapter 5: solvent effects in solvolytic reactions, such as the S_N1 reactions of alkyl halides, and we shall see that the role of the solvent cannot be neglected in the wider aspects dealt with in the present Chapter.

Two main topics are usually distinguished: the influence of the catalyst in proton transfer reactions showing general acid or general base catalysis (the Brönsted equation) and the nucleophilicity of reagents in bimolecular reactions. These topics, however, are not totally unconnected. Although there have been some remarkable advances in the past decade, the application of correlation analysis in this area is still in a relatively elementary state of development. Some qualitative discussion will therefore be found appropriate.

6.2. ACID—BASE CATALYSIS: THE BRÖNSTED EQUATION

The catalysis of organic reactions by acids and bases is a very complicated subject. For detailed accounts appropriate monographs must be consulted, *e.g.* those by Bell[1,2] and by Caldin and Gold.[3] The subject is also dealt with extensively in many monographs on physical organic chemistry, *e.g.* those by Hine,[4] Hammett,[5] and Lowry and Richardson.[6] We shall develop the subject here in a considerably simplified way, as necessary background for the application of correlation analysis.

6.2.1. SOME FORMS OF ACID CATALYSIS

The simplest situation is represented by the following scheme, where S is the substrate and H^+ is the hydrogen ion in a protic

Type (*a*)

$$S + H^+ \overset{K}{\rightleftharpoons} SH^+ \qquad (6.1)$$

$$SH^+ \xrightarrow{k_1} \text{products} \qquad (6.2)$$

solvent, *e.g.* H_3O^+ in water. Equation (6.1) is the equilibrium for rapid reversible protonation of the substrate, while (6.2) is the ensuing rate-limiting conversion of the protonated form into products. The reaction velocity, v, is given by equations (6.3) and (6.4) where k_a is the catalytic constant. This behaviour is essentially shown

$$v = k_1K[H^+][S] \qquad (6.3)$$

$$k_a = k_1K \qquad (6.4)$$

by the hydrolysis of acetals, $RCH(OEt)_2$. Protonation gives $[RCH(OEt)(HOEt)]^+$, which undergoes rate-limiting loss of ethanol to give $[RCH(OEt)]^+$; the latter undergoes a series of very rapid changes leading finally to RCHO.[7] Acid catalysis of this type is often described as *specific* hydrogen ion catalysis.

If the substrate has an ionizable proton, this may become involved in the reaction, and a slightly more complicated scheme may apply.

Type (*b*)
$$\mathit{H}\mathrm{S} + \mathrm{H}^+ \overset{K}{\rightleftharpoons} \mathit{H}\overset{+}{\mathrm{S}}\mathrm{H} \tag{6.5}$$
$$\mathit{H}\overset{+}{\mathrm{S}}\mathrm{H} + \mathrm{A}^- \xrightarrow{k_1} \mathit{H}\mathrm{A} + \mathrm{SH} \tag{6.6}$$

HS is the substrate, and A^- and HA are the components of a buffer solution controlling $[H^+]$. (The italic type is to distinguish the ionizable proton of the substrate from the hydrogen ions in solution.) Through (6.5) and (6.6) HS is converted into an isomeric species SH, which can then undergo reaction with other species present. If such a reaction is very fast, the observed overall rate of reaction will be given by equations (6.7) to (6.10):

$$v = k_1K[\mathrm{HS}][\mathrm{H}^+][\mathrm{A}^-] \tag{6.7}$$
$$\text{or} \quad v = k_1KK_a[\mathrm{HS}][\mathrm{HA}] \tag{6.8}$$
$$\text{where} \quad K_a = [\mathrm{H}^+][\mathrm{A}^-]/[\mathrm{HA}] \tag{6.9}$$
$$k_a = k_1KK_a \tag{6.10}$$

and k_a is the catalytic constant for catalysis by the acid HA. Note, however, that the actual processes involve H^+ and A^-. In principle HA may be any (Brönsted) acid, so this scheme is for *general acid catalysis*.

Processes of the above type certainly play a part in the halogenation (Br_2 or I_2) of acetone in acidic buffer solutions,[8] the intermediate formation of the enol being involved, as shown in equations (6.11) to (6.13).

$$\mathrm{CH_3COCH_3} + \mathrm{H}^+ \underset{\text{fast}}{\overset{\text{very}}{\rightleftharpoons}} \mathrm{CH_3C(OH)^+CH_3} \tag{6.11}$$
$$\mathrm{CH_3C(OH)^+CH_3} + \mathrm{A}^- \xrightarrow{\text{slow}} \mathrm{CH_3C(OH){:}CH_2} + \mathrm{HA} \tag{6.12}$$

$$CH_3C(OH){:}CH_2 \;+\; X_2 \xrightarrow[\text{fast}]{\text{very}} CH_3COCH_2X \;+\; H^+ \;+\; X^- \qquad (6.13)$$

It is also possible for HA to be involved directly in attacking the substrate.

Type (*c*)

$$S \;+\; HA \xrightarrow{k_1} SHA \qquad (6.14)$$

$$SHA \xrightarrow{\text{fast}} \text{products} \qquad (6.15)$$

The observed rate of reaction is that of step (6.14) and is given by (6.16), the catalytic constant k_a being k_1.

$$v \;=\; k_1[S][HA] \qquad (6.16)$$

The acid-catalysed decomposition of diazodiphenylmethane in alcoholic solution is of this type. (This reaction has been mentioned several times in earlier Chapters; *e.g.* Section 5.5.)[9,10]

$$Ph_2CN_2 \;+\; RCO_2H \xrightarrow{\text{slow}} [Ph_2\overset{+}{C}H{\cdot}N_2{-}{-}{-}\bar{O}_2CR] \qquad (6.17)$$

ion pair

$$[Ph\overset{+}{C}H{\cdot}N_2{-}{-}{-}\bar{O}_2CR] \xrightarrow[\text{fast}]{\text{very}} Ph_2\overset{+}{C}H \;+\; N_2 \;+\; RCO_2^- \qquad (6.18)$$

The diphenylmethyl cation very rapidly reacts further, either with RCO_2^- to give the diphenylmethyl ester of the acid or with the solvent alcohol R^1OH to give the ether, Ph_2CHOR^1.†

6.2.2. THE BRÖNSTED EQUATION

A survey of reactions which show general acid catalysis reveals that the catalytic constant, k_a, is related to the strength of the catalysing acid: the stronger the acid, the better it is as a catalyst. A qualitative connection between k_a and K_a, the dissociation constant of the acid, is reasonable in so far as k_a is concerned with ease of proton transfer to the substrate, and K_a measures the extent of proton transfer to water.

$$HA \;+\; H_2O \rightleftharpoons A^- \;+\; H_3O^+ \qquad (6.19)$$

† In so far as carboxylic acid is used up in this system it is not strictly an example of 'catalysis'. The role of proton transfer is, however, the central feature of interest in (*c*), as in (*a*) and (*b*).

For many reactions and acid catalysts, however, a precise quantitative relationship is well obeyed. Following Brönsted and Pedersen (1924)[11] this relationship may be written in the form:

$$k_a = G_a K_a^{\alpha} \qquad (6.20)$$

where G_a and α are constants characteristic of the reaction, the solvent, and the temperature. Values of α usually lie in the range 0.3—0.9. The Brönsted equation may also be written in the form:

$$\log k_a = \log G_a + \alpha \log K_a \qquad (6.21)$$

It is then obviously a linear free-energy relationship, of which it was probably the earliest example.

Some reactions are catalysed specifically by hydroxide ions, but many show *general base catalysis*; see Bell.[1,2] For these the Brönsted equation is written as:

$$k_b = G_b K_b^{\beta} \qquad (6.22)$$

where K_b is given by equation (6.24) for the equilibrium in equation (6.23).

$$B + H_2O \rightleftharpoons BH^+ + OH^- \qquad (6.23)$$

$$K_b = [BH^+][OH^-]/[B] \qquad (6.24)$$

Equation (6.22) was developed by Brönsted and Pedersen[11] in their work on the base-catalysed decomposition of nitramide:†

$$NH_2NO_2 \longrightarrow N_2O + H_2O \qquad (6.25)$$

The Brönsted equation has been successfully applied to many reactions showing acid or base catalysis. It is fairly obvious why type (*c*) processes should conform to the equation, for HA attacks the

† In this Section it has been convenient to concentrate on general acid catalysis and present the Brönsted equation in its form for acid catalysis first, and for base catalysis second. Historically the order was the other way round: Brönsted and Pedersen[11] treated general base catalysis in 1924. The equation was generalized by Brönsted to include acid catalysis in 1926.[12]

substrate directly. It is less obvious for type (*b*), for which $k_a = k_1KK_a$ and k_1 is for attack of $H\overset{+}{S}H$ by A^-. Conformity to the Brönsted equation implies that equation (6.26) holds.

$$k_1KK_a = G_aK_a^{\alpha} \tag{6.26}$$

Since K and G_a are characteristic of the reaction (substrate), solvent, and temperature, it follows that k_1 is proportional to $K_a^{(\alpha-1)}$. Since $\alpha < 1$, this means that k_1 must decrease as K_a increases, *i.e.* the stronger the acid, the poorer is its anion at removing a proton from $H\overset{+}{S}H$, which seems reasonable.

When acid catalysts with more than one acidic centre are included in Brönsted correlations, a 'statistical correction' is required.[1,2] Let us consider an acid with two equivalent acidic centres, *e.g.* succinic acid. There is a statistical factor of two which assists both the catalytic action and the ionization of the acid. In order to fit such an acid into a Brönsted correlation along with catalysts possessing only one acidic centre, k_a and K_a must each be reckoned *per acidic centre*, *i.e.* $k_a/2$ and $K_a/2$ must be used for the acid with two equivalent acidic centres. A related problem arises for the monoanion of the dibasic acid. There are more difficult problems if the acidic centres are not equivalent or if more than one ionizable hydrogen is attached to the same atom. The statistical effect also arises in base catalysis.[1,2]

We now examine a few of the systems to which the Brönsted equation has been applied.

Reference has already been made in Section 4.2.1 (Figure 4.1) to the studies of the acid-catalysed dehydration of acetaldehyde hydrate in aqueous acetone made by Bell and Higginson (1949).[13] This reaction still provides one of the best examples of the application of the Brönsted equation. Some of the results are plotted in Fig. 6.1. Various carboxylic acids, phenols, and NH acids as catalysts conform well to a linear relationship over about ten units of pK_a and five units of $\log k_a$. Acids of other types show marked deviations, some above and some below the line. Thus in Fig. 6.1 the three points above the line at the left of the graph are for two oximes and chloral hydrate, while the five well below the line at the left are for four aliphatic nitro-compounds and benzoylacetone, $PhCO{\cdot}CH_2CO{\cdot}Me$. Deviations from Brönsted plots may be discussed in terms of electronic effects and sometimes of steric effects.[14]

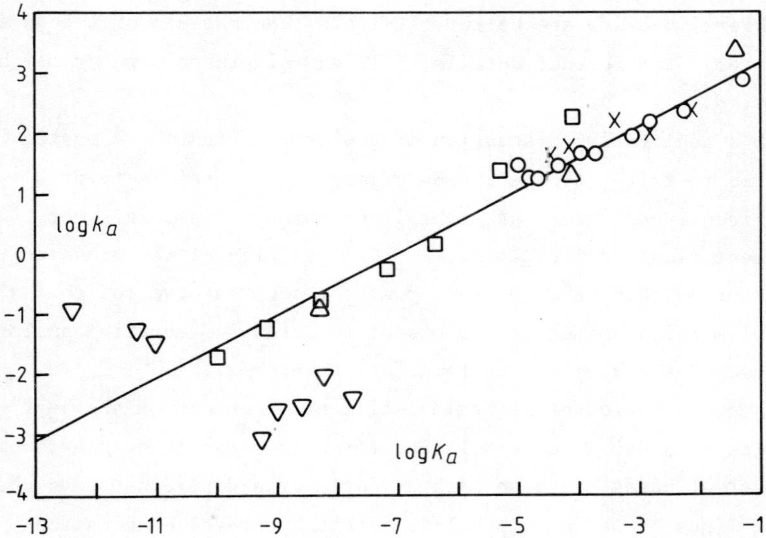

FIG. 6.1. Brönsted plot for acid-catalysed dehydration of acetaldehyde hydrate. Data from Bell and Higginson;[13] k_a in l mol^{-1} min^{-1}, 25°C. Statistical corrections have been applied where necessary. Catalysts are as follows:

○ Aliphatic carboxylic acids. □ Phenols.

X Benzoic acids. △ NH acids.

▽ Miscellaneous acids including oximes and aliphatic nitro-compounds.

To explain the above-mentioned deviations, we need to focus attention on the extent of charge delocalization in the anion of the acid catalyst. Delocalization is much less in the anions from oximes and chloral hydrate than in the anions (carboxylate and phenolate) of the acids defining the line. There is, however, more extensive charge delocalization in the anions of aliphatic nitro-compounds and benzoylacetone. Charge delocalization influences K_a more than k_a, because in the transition state for the acid-catalysed reaction, the anion is not fully formed. In consequence aliphatic nitro-compounds and benzoylacetone are worse catalysts, and oximes and chloral hydrate are better catalysts than phenols of the same strength. For further details, Bell and Higginson's paper may be consulted.[13]

Such a qualitative discussion of deviations from the Brönsted equation is fairly typical. Sometimes it is possible to draw significantly different straight lines through the points for different classes of catalyst.[14] The refining of the Brönsted treatment in this way, however, does not seem to have led to any general multiparameter extensions of the Brönsted equation analogous to those of the Hammett equation (*cf.* Chapter 3).

The decomposition of nitramide still provides one of the best examples of a Brönsted correlation for a base-catalysed reaction.[1,11] For neutral bases or those with a single or a double negative charge, a good linear relationship holds, although careful examination suggests that separate lines might be drawn for each class of base. A very different line is followed by metal complexes which act as base catalysts. Some tertiary amines show marked deviations from a line based on ring-substituted anilines: the tertiary amines are much more effective catalysts than other bases of similar pK_b values.

The Brönsted exponents α and β are often taken to indicate the extent of proton transfer in the activated complex.[14] For this purpose K_a and k_a (or K_b and k_b) should relate to the same solvent and temperature. For the reaction between diazodiphenylmethane and carboxylic acids in ethanol, $\alpha \approx 0.48$.† This may be taken to mean that the proton is about half transferred from the carboxy group to

† This is the ratio of the Hammett ρ values for the reaction of substituted benzoic acids with DDM in ethanol (0.94)[15] and for their ionization in ethanol (1.96).[16]

the central carbon atom of the DDM in the activated complex. However, this type of interpretation is often an over-simplification: other indicators of the symmetry of the activated complex, especially the study of kinetic isotope effects, do not always lead to the same conclusions as those inferred from Brönsted exponents. Further there are 'anomalous' Brönsted exponents, *i.e.* values greater than unity or negative in sign, which are unintelligible on the basis of 'extent of proton transfer'.[14]

In recent years the application of fast-reaction techniques[17] has greatly extended our knowledge of proton transfer processes, and has revealed a limitation of Brönsted correlations which is probably general. It seems likely that all Brönsted plots would be seen as curves if catalysts with a sufficiently wide range of pK_a or pK_b value could be studied.[14]

The process in equation (6.27) has been studied for catalysts covering about seventeen units of pK_a.[18]

$$[CH_3CO{\cdot}CHCO{\cdot}CH_3]^- + HA \underset{k_2}{\overset{k_1}{\rightleftharpoons}} CH_3CO{\cdot}CH_2CO{\cdot}CH_3 + A^- \qquad (6.27)$$

The plot of log k_1 *versus* pK_a is shown in Fig. 6.2 (*see page 182*). For carboxylic acids or phenols as catalysts a straight line could be drawn with $\alpha \approx 0.4$ but a wider range of catalysts reveals marked curvature, with α tending to unity at one end and zero at the other. The detailed explanation of this and similar behaviour is quite complicated, but the flattening at one end ($\alpha \to 0$) is connected with the tendency of the process to become diffusion-controlled when HA is a rather strong acid. For very weak acids the reverse reaction tends to become diffusion-controlled, and k_1 becomes proportional to the equilibrium constant for the process, and hence to the ordinary dissociation constant of the acid HA ($\alpha = 1.0$).

The above account of acid—base catalysis must suffice within the limitations of the present book. The references given will supply ample guidance for fruitful further reading. However, the above account provides the background necessary for the treatment of nucleophilicity in bimolecular reactions, to which we now come.

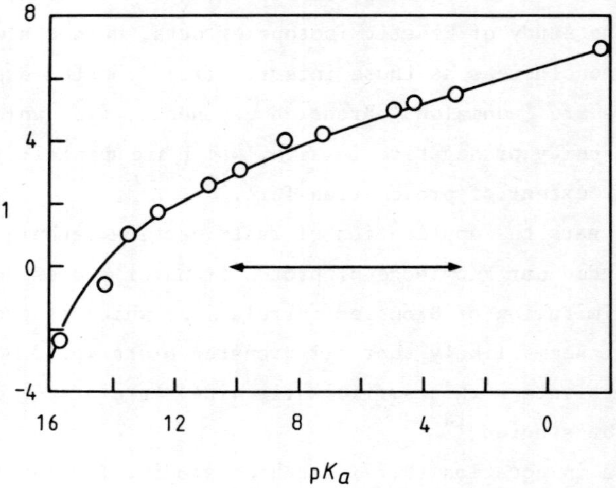

FIG. 6.2. Brönsted plot for transfer of proton to enol of acetylacetone. Data from Eigen.[18] k_1 in $1\ mol^{-1}\ sec^{-1}$, 27.5°C. The arrows indicate the approximate pK_a range of phenol and carboxylic acid catalysts.

6.3. NUCLEOPHILICITY IN BIMOLECULAR REACTIONS[19,20]

6.3.1. BRÖNSTED-TYPE RELATIONSHIPS

A nucleophilic displacement reaction may be regarded essentially as an acid—base reaction; the nucleophile (base, electron-donor) attacks the substrate (Lewis acid, electron-acceptor). In the light of the success of the Brönsted equation in proton transfer processes, some relationship might be expected between the ability of an anion A^- to act as a nucleophile in displacement reactions and the pK_a value of the acid HA; the best nucleophiles should be the anions of the weakest acids. For an uncharged nucleophile B there should be a relationship with its pK_b value or with the pK_a value of the conjugate acid BH^+.

There are indeed many examples in which log k for a nucleophilic displacement reaction may be correlated with the pK_a value of the conjugate acid of the nucleophile through equation (6.28):

$$\log k = \beta_{nuc} pK_a + \text{const.} \qquad (6.28)$$

The simplest cases involve varying the structure by introducing fairly remote substituents into the moiety that is attached to a given nucleophilic centre, *e.g.* the nucleophiles could be a series of *meta*- or *para*-substituted phenoxide ions.[21] In such a case conformity to equation (6.28) follows from conformity to the Hammett equation (with an appropriate set of substituent constants) or other simple LFER, on the part of both the processes involved. Correlation analysis for the reactions of such a homogeneous set of nucleophiles presents no problems essentially different from those of any other reactions which may be considered in connection with the Hammett equation and its extensions, Taft's equations, *etc.* (see Chapters 3 and 4). In some cases, however, equation (6.28) may embrace data which cannot be fitted into simple individual LFER, *e.g.* log k values for various ring-substituted anilines reacting with benzoyl chloride in benzene solution are linearly related to pK_a values for the anilinium ions in aqueous solution, and the linear relationship includes *ortho*-substituted anilines.[22]

The correlation analysis of nucleophilicity becomes much more of a problem when a wide range of nucleophiles is involved, as in the Brönsted plot Fig. 6.3 for the hydrolysis of *p*-nitrophenyl acetate initiated by nucleophilic attack on the carbonyl group of the ester.[23]

$$p\text{-}NO_2C_6H_4OCOCH_3 + B \xrightarrow{\text{slow}} CH_3COB^+ + p\text{-}NO_2C_6H_4O^- \qquad (6.29)$$

$$CH_3COB^+ + H_2O \xrightarrow{\text{fast}} CH_3CO_2H + BH^+ \qquad (6.30)$$

There is indeed approximate conformity to equation (6.28) and the slope of the line is about 0.8, which is the value of β_{nuc}. But some nucleophiles, such as OH^-, react much more slowly than expected, while others such as N_3^- react more rapidly. We have already noted the same kind of scatter when a great variety of acidic or basic catalysts is used in an ordinary Brönsted plot for proton-transfer processes

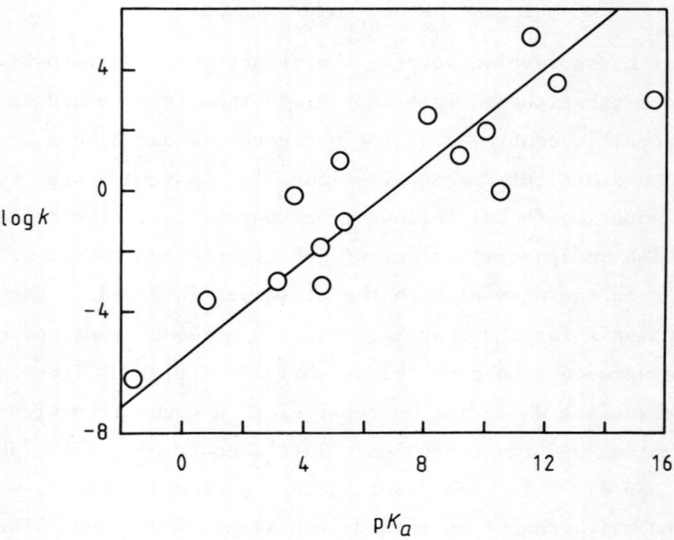

FIG. 6.3. Attempted Brönsted-type plot for nucleophilic attack on *p*-nitrophenyl acetate. After Jencks and Carriulo.[23] k in $1\ mol^{-1}\ min^{-1}$, 25^oC; pK_a values are for the conjugate acids; no statistical corrections have been applied, but the plot would not be much changed if they were. Nucleophiles were a variety of inorganic and organic bases, involving various donor atoms; the species were variously neutral or charged (±).

(Section 6.2.2). Further, it is easy to find nucleophilic displacement reactions for which any connection between nucleophilicity and pK is tenuous in the extreme. Such a situation is presented by the attack of nucleophiles on methyl iodide in methanol;[24] a plot of log k *versus* pK_a values for about thirty nucleophiles shows an almost random distribution of points.

In spite of their limitations, Brönsted-type plots have often proved valuable as a basis for discussing nucleophilic displacement reactions. Duboc[20] summarized many examples of this, taken from the literature to about 1975. She also includes various examples of extensions of equation (6.28). Thus equation (6.31) may be applied to equilibria for the addition of nucleophiles to electrophilic

$$\log K = \delta \mathrm{p}K_a + \text{const.} \qquad (6.31)$$

centres (cationic or neutral), *e.g.* the formation of Meisenheimer-type complexes, as in the addition of substituted thiophenoxides to 1,3,5-trinitrobenzene.[25] The equilibrium constants K for processes such as the latter (carbon basicity) may also provide a standard which is sometimes more satisfactory than that provided by pK_a values in the Brönsted-type correlation analysis of nucleophilicity, *i.e.* equation (6.32) applies.[26]

$$\log k = \eta \log K + \text{const.} \qquad (6.32)$$

We conclude this Section with a few examples of Brönsted-type correlations of nucleophilicity taken from the more recent literature.

Bordwell and Hughes[27] have measured the rates of reaction for five families of 9-substituted fluorenyl anions, 9-G-Fl^- (G = CO_2Me, Me, $PhSO_2$, PhS, or Ph) reacting with benzyl chloride in DMSO. The CO_2Me and Me families had substituents in the 2- or 2,7-position(s) of the fluorene ring; the remaining families carried substituents in the *meta*- or *para*-position of the Ph ring. Thus each family constituted a series of nucleophiles of constant steric effect. Brönsted-type plots of log k *versus* pK_a for these five families gave parallel lines of slopes $\beta_{nuc} \approx 0.3$. From the relative positions of the lines the steric effect of G in 9-G-Fl^- in these S_N2 reactions is seen to increase in the order: $CO_2Me < Me < PhS < PhSO_2 < Ph$. The paper contains much detailed discussion of the above and related results.

Rate constants for nucleophilic reaction of nine substituted pyridines with methyl chloroformate give a markedly curved Brönsted-type plot, with β_{nuc} changing from 0.93 to 0.15 with increasing reactivity.[29] This contrasts with the straight line, $\beta_{nuc} = 0.93$, observed for the pyridinolysis of *p*-nitrophenyl acetate, and is shown to be quantitatively consistent with a change in rate-determining step.

We have already referred in this Section to the use of a Brönsted-type plot in the work of Jencks and Carriulo (1960); Fig. 6.3.[23] Such plots have played a major role in later work of Jencks and his colleagues; see, for example, their work on acyl transfer between sulphur and oxygen nucleophiles,[29] and on the effect of solvation on β_{nuc} values.[30]

6.3.2. THE SWAIN—SCOTT EQUATION

The limitations of the Brönsted-type correlation analysis of nucleophilicity suggest that the development of a more effective procedure for nucleophilic displacement reactions should involve operating, initially at least, within the system of such reactions alone, without reference to the basicity of nucleophiles as measured by pK_a values. Swain and Scott (1953)[31] suggested an LFER for nucleophilicity:

$$\log (k/k^{o}) = sn \tag{6.33}$$

where k is the rate constant for reaction of a certain nucleophile with a given substrate, k^{o} is for the reaction of the same substrate with water, n is a nucleophilicity constant (analogous to σ) and s is a susceptibility constant (analogous to ρ). Methyl bromide was taken as the standard substrate (s = 1.00, 25°C); n for water was set equal to 0.00, at 25°C.

Selected values of n are given in Table 6.1, along with the pK_a values of the conjugate acids, and are discussed below. The Swain—Scott equation may be used to correlate the results of a variety of bimolecular nucleophilic displacement reactions, as shown in Fig. 6.4. The data in this diagram are from the 1953 paper. Some scatter of points is apparent and thus, from its inception, the Swain—Scott equation was clearly only approximate in character. It is perhaps an indication of the difficulties of this area of correlation analysis that the original Swain—Scott equation is still found to be of value, almost thirty years later. Thus Biggs and Williams have shown that the rates of nucleophile-catalysed denitrosation of *N*-methyl-*N*-nitrosaniline by Cl^-, Br^-, I^-, and SCN^- are well correlated by the Swain—Scott n values.[32] The Swain—Scott equation has also been applied recently to the second order reactions of nucleophiles with the acetal 1-methoxymethoxy-2,4-dinitrobenzene,[33] and to the rates of aminolysis of α-chloropropionic acid by aliphatic amines.[34]

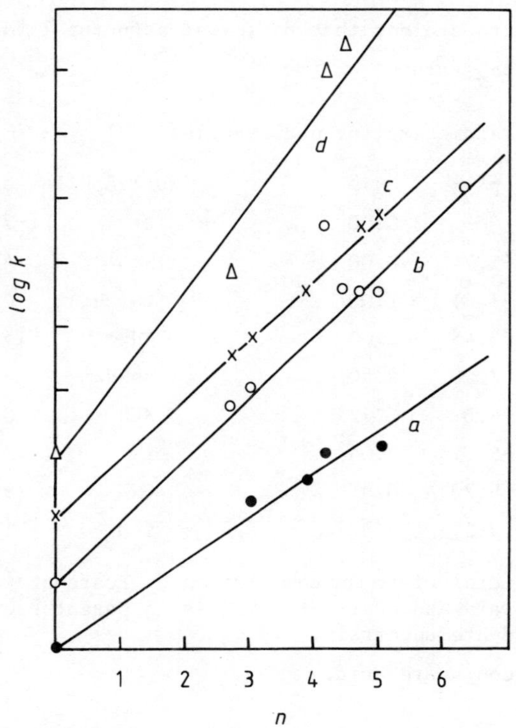

FIG. 6.4. Application of the Swain—Scott equation. Data from Swain and Scott.[31] The divisions on the ordinate are 1.00 unit of log k apart, and the relative positions of the lines a to d with respect to the ordinate are arbitrary. The substrates are: a. ethyl tosylate; b. mustard cation; c. epichlorohydrin; d. benzoyl chloride. See the original paper for further details. Nucleophiles may be identified from n-values in Table 6.1.

We return now to Table 6.1. There is little connection between n and pK_a values overall, but some connection can be seen within limited ranges of nucleophiles. Thus the anions of the strongest oxyacids are poor nucleophiles, and the order $ClO_4^- < NO_3^- < SO_4^{2-} < OAc^-$

< HPO_4^{2-} < OH^- is reasonable in relation to the pK_a values. The halide ions, however, are in the order I^- > Br^- > Cl^- > F^- and so I^- (from the strong acid HI) is a better nucleophile than F^- (from the weak acid HF) and is also better than OH^- (from the very weak acid H_2O). It is thus clear that different structural factors control n values and pK_a values.

TABLE 6.1. Parameters for nucleophiles[a]

Nucleophile	pK_a[b]	n	Nucleophile	pK_a[b]	n
ClO_4^-	-	<0.00	Br^-	(-7.7)	3.89
H_2O	-1.74	0.00	N_3^-	4.74	4.00
NO_3^-	(-1.3)	1.03	$CS(NH_2)_2$	-0.96	4.1
F^-	3.45	2.00	OH^-	15.7	4.20
SO_4^{2-}	2.00	2.50	$PhNH_2$	4.6	4.49
OAc^-	4.8	2.72	SCN^-	(-0.7)	4.77
Cl^-	(-5.7)	3.04	I^-	(-10.7)	5.04
C_5H_5N	5.23	3.6	CN^-	9.3	5.1
HPO_4^{2-}	7.2	3.8	$S_2O_3^{2-}$	1.9	6.36

a pK_a values mainly from the compilation by Pearson.[19] n values mainly from Swain and Scott.[31] Values in parentheses are estimated indirectly and are uncertain.

b pK_a of the conjugate acid.

An obvious factor is ionic radius: large anions tend to be more reactive than small ones, because large anions, being polarizable, are able to minimize electrostatic repulsion in the trigonal-bipyramidal activated complex, I.

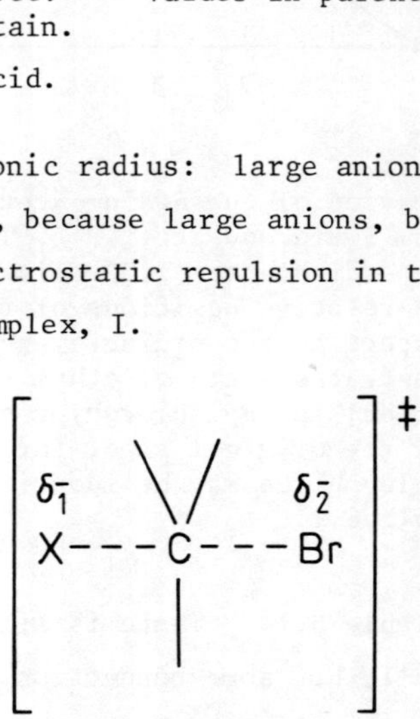

(I)

Polyatomic anions, *e.g.* SCN^- and $S_2O_3^{2-}$, are especially effective in this respect, since the partial negative charge can be distributed over several atoms. There is also no doubt that ionic solvation is often a very important factor. We have already referred to the solvation of anions in hydroxylic solvents as important for solvent effects on nucleophilic substitutions (Section 5.1.3). It seems probable that the order of n values in aqueous solution $I^- > Br^- > Cl^-$ is very largely due to solvation in the order $Cl^- > Br^- > I^-$ having to be broken down in the formation of the activated complex. In the absence of such solvation, *e.g.* in acetone solution, the order of nucleophilicities is $Cl^- > Br^- > I^-$.

The role of solvation in governing nucleophilicity has recently been strikingly confirmed in various ways. Thus Young *et al.*[35] have found that the nucleophilic reactivities of H^-, F^-, OH^- and NH_2^- towards CH_3Cl in the gas phase at 300K are approximately the same and do not correlate with gas phase basicities. This suggests that the large variations in the reactivities of the anions in solution are to be attributed to solvation. Further, Liotta *et al.* have studied the reactivities of 'naked' anions in solution.[36] This condition is achieved through the solubilization of their potassium salts in acetonitrile by means of 18-crown-6. In reaction with benzyl tosylate under these conditions, at 30°C, F^- is about 1.4 times as reactive as I^-. This contrasts with a difference of about 3 units in n, corresponding to I^- being a thousand times more reactive than F^- in aqueous solution.

Thus the Swain—Scott relation in its original form as presented above is essentially a relationship for aqueous solutions or aqueous organic mixtures. Nevertheless the n values correlate fairly well with the log k values for reactions of nucleophiles with MeI in methanol, the system mentioned in Section 6.3.1 as showing no Brönsted-type relationship. This latter system has been suggested as the basis of an alternative scale of nucleophilicity, n_{MeI}.[19,24] (The original Swain—Scott n is now sometimes distinguished as n_{MeBr}.) Some illustrative values of n_{MeI} are presented in Table 6.2. This scale has been used for correlation analysis in certain studies, *e.g.*

of the reactivities of the cation radical of 9,10-diphenylanthracene towards various nucleophiles.[37]

TABLE 6.2. Further parameters for nucleophiles[a]

Nucleophile	n_{MeI}	n_{Pt}	Nucleophile	n_{MeI}	n_{Pt}
MeOH	0.00	0.00	Br^-	5.79	4.18
NO_3^-	(1.5)	-	OMe^-	6.29	<2.4
F^-	≈2.7	<2.2	NH_2NH_2	6.61	3.86
SO_4^{2-}	(3.5)	-	NMe_3	6.66	-
OAc^-	4.3	<2.0	CN^-	6.70	7.14
Cl^-	4.37	3.04	SCN^-	6.70	5.75
C_5H_5N	5.23	3.19	$CS(NH_2)_2$	7.27	7.17
$PhNMe_2$	5.64	-	I^-	7.42	5.46
$PhNH_2$	5.70	3.16	PEt_3	8.72	8.99
PhSH	5.70	4.15	$S_2O_3^{2-}$	8.95	7.34
PhO^-	5.75	-	PhS^-	9.92	7.17
N_3^-	5.78	3.58			

a From Pearson's compilation.[19] Values in parentheses are estimated indirectly and are uncertain. n_{MeI} and n_{Pt} are both determined for reaction in methanol at 25°C.

It is, however, not difficult to find substrates which present nucleophilic reactivity orders very different from the order of n or n_{MeI} values. Thus the reaction:

$$B + trans\text{-}Ptpy_2Cl_2 \rightarrow trans\text{-}Ptpy_2ClB^+ + Cl^- \quad (6.34)$$

has been suggested as another reference reaction for assessing nucleophilicities (py = pyridyl).[19,24] Selected n_{Pt} values are in Table 6.2. While the n_{MeI} and n_{Pt} scales show a rough parallelism for limited ranges of nucleophiles, there are many discrepancies of detail between the scales. The reactions of several other electrophilic centres, in both organic and inorganic compounds and in various solvents, have also been suggested as reference processes, giving, for example, n_{Pd};[38] and n_P (displacement on phosphorus), n_O (oxygen), n_S (sulphur), and n_{car} (aromatic carbon).[39]

It is clear that the Swain—Scott equation is much too simple to be of wide application, whatever standard reaction may be chosen.

6.3.3. THE EDWARDS EQUATION

Edwards (1954)[40] proposed an equation to express the operation of two factors governing nucleophilicity. The equation takes the form:

$$\log (k/k^{o}) = \alpha E_{n} + \beta H \quad (6.35)$$

where α and β characterize the substrate and E_n and H characterize the nucleophile. The quantity E_n is related to the electrode potential for a redox half-reaction in which the nucleophile is the reduced form, *e.g.*

$$2Br^{-} \rightleftharpoons Br_{2} + 2e^{-} \quad (6.36)$$

The definition of E_n is:

$$E_{n} = E^{o} + 2.60 \quad (6.37)$$

where E^o is the standard potential (American convention) for the half-reaction and the 2.60 is introduced to give $E_n = 0$ for water, E^o for the half-reaction

$$2H_{2}O \rightleftharpoons H_{4}O_{2}^{2+} + 2e^{-} \quad (6.38)$$

being -2.60 volts.

H is defined by the equation

$$H = pK_{a} + 1.74 \quad (6.39)$$

where pK_a is for the conjugate acid of the nucleophile and the 1.74 comes from the notional pK_a value for H_3O^+, -1.74. Selected values of E_n and H are in Table 6.3.

The Edwards equation is essentially a Brönsted equation (the βH term) modified by a term (αE_n) which expresses the polarizability of the nucleophile. E_n is undoubtedly related to the ease of removal of electrons from the nucleophile, which may in turn be held to reflect its polarizability. A large value of β/α thus indicates a Brönsted-type behaviour, while a large value of α/β indicates the dominant role of nucleophile polarizability.

The Edwards equation is quite successful for a variety of displacement reactions but its application is limited by the fairly small number of nucleophiles for which there are electrochemically-determined values of E_n. Other values have been determined by interpolation on Edwards plots (see Table 6.3). Further, the extent

TABLE 6.3. Edwards parameters for nucleophiles[a]

Nucleophile	E_n	H	P
ClO_4^-	<0.00	(-9.0)	-0.046
F^-	-0.27	4.9	-0.150
H_2O	0.00	0.00	0.00
NO_3^-	(0.29)	(0.4)	0.0
SO_4^{2-}	0.59	3.74	0.0035
OAc^-	(0.95)	6.46	-
C_5H_5N	(1.20)	7.04	-
Cl^-	1.24	(-3.0)	0.389
Br^-	1.51	(-6.0)	0.539
OH^-	1.65	17.48	0.143
$PhNH_2$	(1.78)	6.3	0.921
SCN^-	1.83	(1.0)	-
I^-	2.06	(-9.0)	0.718
$CS(NH_2)_2$	2.18	0.08	-
CN^-	2.79	10.88	0.373
S^{2-}	3.08	14.66	0.611

a Taken mainly from the compilation of Ibne-Rasa.[48] Many of the values are those proposed by Edwards.[40,41,49] Values in parentheses were determined indirectly and are uncertain.

to which equation (6.35) really effects a clean separation of nucleophilicity into polarizability and basicity effects must be considered doubtful. In certain cases β was found to be negative, thus implying an inverse relationship of nucleophilicity to basicity, clearly an unacceptable idea. Edwards himself suggested that the explanation of the negative β values was that E_n values contained a significant contribution from basicity. He therefore modified the equation to get over this problem.[41]

It was proposed that the composite nature of the E_n values should be expressed through the equation:

$$E_n = aP + bH \tag{6.40}$$

where H is again as defined in equation (6.39) and P is a measure of the polarizability of the nucleophile relative to that of water, *i.e.*

$$P = \log (R/R_{H_2O}) \tag{6.41}$$

with R standing for molar refractivity at infinite wavelength. a and b are constants whose values are 3.60 and 0.0624 respectively. The combination of equations (6.40) and (6.35) gives equation (6.42):

$$\log (k/k^o) = AP + BH \tag{6.42}$$

in which A and B are new substrate constants. They are related to the substrate constants in equation (6.35) by $A = 3.60\alpha$ and $B = 0.0624\alpha + \beta$. Values of P for selected nucleophiles are shown in Table 6.3. In applications of equation (6.42), B is never negative.

In spite of its limitations, the original Edwards equation (6.35) continues to find applications. See, for example, recent studies of nucleophilic substitution on *ortho*-nitrobenzenesulfenyl and *ortho*-nitrobenzeneselenenyl halides.[42] In recent years the Edwards equation has also been found useful by inorganic chemists in a variety of ways, *e.g.* there have been applications to the effect of ligands on the standard electrode potentials of aqua gold ions,[43] to the equilibria of formation of adducts of Cu(II) macrocyclic tetra-amine complexes with anions,[44] and to the kinetics and equilibria of various ligand-substitution reactions of Cu(II).[45,46]

Finally in this Section we may appropriately mention certain review articles from the 1960s which are still useful: factors determining nucleophilic reactivities (Edwards and Pearson, 1962),[47] equations for correlation of nucleophilic reactivity (Ibne-Rasa, 1967),[48] and bimolecular nucleophilic displacement reactions (Edwards, 1968).[49]

6.3.4. HARD AND SOFT ACIDS AND BASES

It will already be apparent that a comprehensive understanding of nucleophilicity in quantitative terms through correlation analysis presents formidable difficulties, and we have given but a slight indication of the complexity of much of the data. In the circumstances, qualitative discussion is of considerable interest to the extent that it may provide some degree of rationalization for data which are at present outside the scope of quantitative treatment. Much qualitative rationalization is possible through applying the principle of *hard and soft acids and bases* (HSAB). The main ideas of HSAB originated with R. G. Pearson in the 1960s and for a detailed exposition his articles should be consulted.[19,24,50]

Hard bases are those which hold on to their electrons very strongly, are not easily oxidized and are of low polarizability. They mainly involve F, O, or N as the donor atom. Soft bases have the opposite characteristics and involve donor atoms such as P, S, and I. Hard acids are Lewis acids in which the acceptor atom has one or more of the following characteristics: it is small, it is of high positive charge, or it lacks unshared pairs of electrons in the valence shell. Electrophiles such as H^+, Al^{3+}, BF_3, and RCO^+ are hard, whereas CH_3^+, $C_6H_5^+$, I^+, and Pt^{2+} are soft. A fundamental principle of HSAB is that hard bases react preferentially with hard acids and that soft bases do so with soft acids. Mismatched processes, that is soft with hard, occur much less readily. (It should be mentioned that a more detailed classification includes the categories of 'borderline' acids and bases.)

According to the HSAB principle, the parallelism expressed in Brönsted relationships is connected with proton transfer reactions and acid—base equilibria both involving reactions of nucleophiles with the hard electrophilic centre, the proton. Log k values for reactions at other electrophilic centres will correlate with pK_a values provided that the electrophilic centres are hard. This is more or less true for reactions at carbonyl groups (see Section 6.3.1 and Fig. 6.3). When the electrophilic centre is soft, as is the case for the carbon atom of methyl iodide or the Pt(II) atom of *trans*-$Ptpy_2Cl_2$ (see Section 6.3.2), different factors influence reactivity, and correlation with pK_a values is far worse.

These ideas are developed into a very satisfying rationalization of a wide variety of data. In the space available we must restrict ourselves to a few examples.

The order of softness of the halide ions is $I^- > Br^- > Cl^- > F^-$.

In reaction with soft electrophilic centres the order of reactivity tends to be $I^- > Br^- > Cl^- > F^-$, while with harder centres the opposite order may be found. With very hard centres, the larger ions, all of which are fairly soft, may not react at all, although the harder F^- may do so.

These considerations may be extended to include the solvent-dependence of nucleophilicity orders. Protic solvents solvate most strongly the hard anions, and solvation decreases markedly in the following series: $OH^-, F^- \gg Cl^- > Br^- > N_3^- > I^- > SCN^-$. Dipolar aprotic solvents behave as soft acids in their solvation of anions, and interact preferentially with soft anions; in the above series of anions, solvation decreases somewhat in the opposite direction. In a protic solvent the nucleophilicity order with respect to tetrahedral carbon (soft) is:

$$I^- > Br^- > Cl^- > F^-$$

this being the order of decreasing softness. However, in a dipolar aprotic solvent the effect of solvation of the anion opposes that of softness and the nucleophilicity order may be reversed:

$$F^- > Cl^- > Br^- > I^-$$

The physical basis of rationalization in terms of HSAB is not always clear, and HSAB seems to be a convenient way of dealing with a variety of factors. Some progress in the quantitative formulation of HSAB has, however, been made in recent years (see Duboc[20]). There is some relationship to the characteristic quantities of the Edwards equation (Section 6.3.3).

6.3.5. C. D. RITCHIE'S WORK

During the past ten years Ritchie and his colleagues have extensively investigated the reactions between nucleophiles and large organic cations, *e.g.* aryldiazonium, triarylmethyl, and aryl-tropylium cations, in various solvents.[51-3] In many such reactions the combination of the nucleophile and the cation is rate-determining, and these studies thus provide a means of studying nucleophilicity without any complications which may be introduced by the leaving group in ordinary bimolecular nucleophilic substitution.

Ritchie found that the data for many systems of the above type could be correlated by the simple linear free-energy relation:[51]

$$\log k_N = \log k_{H_2O} + N_+ \qquad (6.43)$$

where k_N is the rate constant for reaction of a cation with a given nucleophilic system (*i.e.* a given nucleophile in a given solvent), k_{H_2O} is the rate constant for the reaction of the same cation with water in water, and N_+ is a parameter which is characteristic of the nucleophilic system and independent of the cation. Rate constants for the reactions with solvents are in units of sec^{-1}, while those for reactions with other nucleophiles are in $l\ mol^{-1}\ sec^{-1}$. Much of the earliest work involved anionic nucleophiles but it was soon found possible to include neutral ones such as amines in the same relationships. Equation (6.43) implies that in these reactions relative reactivities of electrophiles are independent of nucleophile, and those of nucleophiles are independent of electrophile.[52]

The N_+ values were defined originally by the rate constants for reactions of '*p*-nitromalachite green' [bis-(*p*-dimethylaminophenyl)-*p*-nitrophenylmethyl cation] wherever possible but reactions with other cations were used for nucleophiles whose reactions with the above substrate were not amenable to study.[51] More recently Ritchie has used a statistical evaluation of N_+ values, starting from a basic data set of 204 rate constants from the reactions of 37 nucleophilic systems (prime nucleophiles) and 23 electrophiles.[54] Selected values of N_+ are in Table 6.4,[54] and Fig. 6.5 shows a typical Ritchie plot.[55] In the course of his early work in this area Ritchie suggested that the N_+ value in some way reflects the solvation of the nucleophiles, or more precisely the extent to which the nucleophile must be desolvated in passage to the transition state. Certainly the high values of N_+ for a given nucleophile relate to solvents in which the nucleophile may be presumed to be relatively weakly solvated. However, detailed understanding of this and of the way in which N_+ values for a range of nucleophiles vary with a given solvent has not so far been achieved.[52,54]

The scope of the application of N_+ values through equation (6.43) has, however, increased enormously: it is no longer restricted to

TABLE 6.4. Ritchie's N_+ parameter[a]

Nucleophile	(solvent)	N_+	Nucleophile	(solvent)	N_+
H_2O	(H_2O)	0.0	OMe^-	(H_2O)	7.28
MeOH	(MeOH)	1.18	OMe^-	(MeOH)	7.68
CN^-	(H_2O)	3.67	CN^-	(DMSO)	8.60
p-$NO_2C_6H_4O^-$	(H_2O)	4.05	N_3^-	(MeOH)	8.85
$PhNH_2$	(H_2O)	4.10	N_3^-	(DMSO)	10.07
OH^-	(H_2O)	4.75	PhS^-	(MeOH)	10.51
C_5H_5N	(H_2O)	5.00	PhS^-	(DMSO)	12.83
PhO^-	(H_2O)	5.6	F^-	(H_2O)	>3.45
CN^-	(MeOH)	5.94	OAc^-	(H_2O)	>2.95

a From ref. 54.

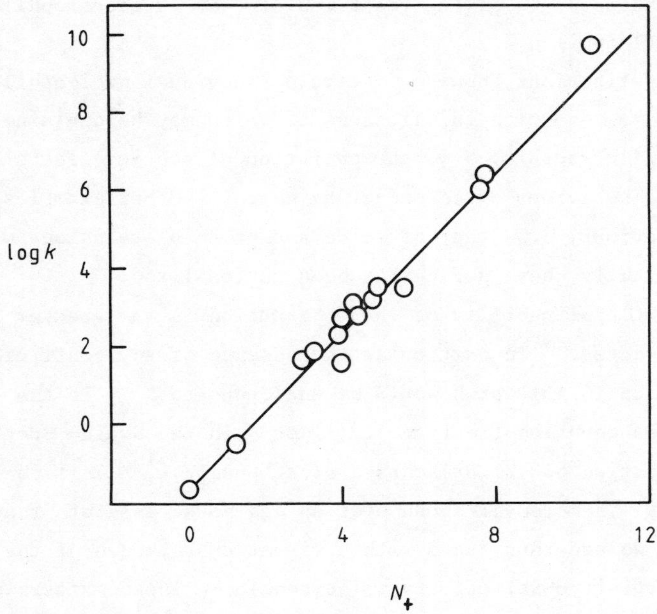

FIG. 6.5. Ritchie plot for the reaction of p-dimethylaminophenyltropylium ion, with a variety of nucleophiles in water or methanol. k in 1 mol^{-1} sec^{-1}, or sec^{-1} (as appropriate), 23°C. After Ritchie and Virtanen.[55]

cation-nucleophile reactions. There are fairly straightforward applications to nucleophilic aromatic substitution in protic solvents[53] and to nucleophilic substitution reactions of very reactive esters; in all these cases the attack of the nucleophile on the substrate is rate-determining. More remarkable, however, are applications to the reactions of nucleophiles with esters which require explicit consideration of a two-step mechanism.[55] The observed rate constants for these reactions, in which the combination step is not always rate-determining, require correction for the relative leaving-group abilities of the nucleophiles. Ritchie was able to assign a single set of leaving-group abilities and the corrected rate constants, *i.e.* now relating to the initial attack of the nucleophile, gave very good correlations. Ritchie claimed in 1978[52] that equation (6.43) had been applied successfully to nearly 800 rate constants, involving reactions of over 30 electrophiles with about 80 nucleophiles.

In many applications, however, certain individual nucleophiles show deviations from equation (6.43), some of which may be explained by highly specific factors, *e.g.* the operation of steric effects in the nucleophilic behaviour of secondary amines.[52] Other examples of deviant behaviour, *e.g.* that of azide and of thiolate anions which occurs frequently, have not so far been rationalized.[52]

The range of applicability of the N_+ equation is far greater than could be expected. In particular the absence of any coefficient of N_+ in equation (6.43) which would be analogous to β_{nuc} in the Brönsted-type equation (Section 6.3.1) or s in the Swain—Scott equation (Section 6.3.2) presents a great puzzle.[51] This constant 'selectivity' is moreover shown over an enormous range of 'reactivity', and we are thus faced with a gigantic breakdown of the often assumed reactivity—selectivity relationship. These matters have been discussed but have not really been completely resolved.[56,57]

Certain further applications of N_+ have involved generalizing equation (6.43) in the form of (6.44),

$$\log k_N = \phi N_+ + B \qquad (6.44)$$

where B is the intercept corresponding to log k_{H_2O} and ϕ is characteristic of the electrophile. (In all the applications considered above $\phi = 1.00$.) The best established application of (6.44) is to the reactions of various amines and some anions with ethylene oxide.[58] A logarithmic plot of rate constant against the pK_a values of the conjugate acids of the nucleophiles gives a curve with certain seriously deviating points, but against N_+ a reasonable straight line is obtained. The value of $\phi \approx 0.6$, and there are two or three seriously deviating points in 15. According to the authors "a slope of less than unity directs attention to the changes occurring in ethylene oxide itself during the formation of the transition state".

Certain other authors have also discussed the possibility of $\phi \neq 1.00$.[59,60] Kice *et al.*[61,62] have had variable success in applying N_+ to their extensive results on nucleophilic substitution at sulphur in various oxidation states.

6.3.6. JENCKS'S EQUATION

This may be regarded as a development in the area of nucleophilic addition to carbonyl compounds which is somewhat analogous to Ritchie's equation (Section 6.3.5) for the combination of nucleophiles with organic cations. Sander and Jencks[63] proposed to express the equilibrium constant for the addition of various nucleophiles to aldehydic carbonyl groups through equation (6.45).

$$\log K_o = \Delta\gamma + A \qquad (6.45)$$

In this equation Δ is a measure of the sensitivity of the carbonyl compound to the affinity of nucleophilic reagents and γ is a measure of this affinity for a given nucleophile. A scale of γ was defined through the equation:

$$\gamma = \log K_{HX}/K_{MeNH_2} \qquad (6.46)$$

where K_{HX} and K_{MeNH_2} are the respective equilibrium constants for the addition of the nucleophile HX and of monomethylamine to the reference carbonyl group in pyridine-4-carboxaldehyde (for which Δ was defined as unity).

For additions to formaldehyde and to *p*-chlorobenzaldehyde Sander and Jencks[63] found good correlations with γ, and the correlation lines were of unit slope, *i.e.* $\Delta \approx 1.00$, even though at a given value of γ, the K_o values for the systems covered a range of 6 or 7 powers of 10, *cf.* the situation for N_+ correlations.

The γ values have subsequently been found to correlate K_o values, with $\Delta = 1.00$, for the addition of nucleophiles to C=N in the quinazolinium cation.[64]

Slopes approximating to unity have also been found when the γ scale is used to correlate equilibrium data for nucleophilic addition to C=C in 1,3-dimethyl-5-nitrouracil;[65] also to C=O in acetaldehyde,[66] pyruvamide,[67] and several other carbonyl compounds.[68]

Very recently a Ritchie-type or Jencks-type correlation expression has been found for rates of nucleophilic addition to carbonyl groups.[69] For a series of 26 ketones, the reactivity towards various nucleophiles (BH_4^-, CN^-, SO_3^{2-}, NH_2OH, RS^-) conforms to an equation of the form:

$$\log k = \log k_o + B \tag{6.47}$$

where k is the rate constant for a ketone reacting with a given nucleophile, and k_o is that for cyclohexanone reacting with the same nucleophile, *i.e.* the reactivity of a given ketone relative to that of cyclohexanone is independent of the nucleophile chosen. The ketones vary in reactivity by about 13 powers of ten, and the nucleophiles by about 6 powers of ten. It is also claimed that the same type of relationship holds for 'reverse reactions', *i.e.* in carbinolamine decomposition, chromic acid oxidation, and decomposition of bisulphite addition compounds, relative reactivity depends only on the ketone, and not on the reaction.

6.3.7. MISCELLANEOUS MATTERS AND CONCLUSION

The foregoing Sections have served to show why it was asserted in Section 6.1 that "the application of correlation analysis in this area is still in a relatively elementary state of development". The situation is analogous to that which existed in the area of solvent effects for many years, *i.e.* when the emphasis lay on attempts to find simple linear relationships of log k with 'solvent polarity' scales, and to explain qualitatively such deviations as were found (Chapter 5). For solvent effects the break-through came when the roles of the various aspects of solvent-solute interaction were more clearly distinguished and the application of multiple regression became possible. Hopefully we may look for an analogous break-through in the area of nucleophilicity but the problems seem immense.

In the foregoing discussion we have mentioned at various times three principal factors influencing nucleophilicity: basicity, polarizability, and solvation, but except in Edwards's work no distinctive parametrization of any of these may reasonably be claimed (Section 6.3.3). The limited effectiveness of any one scale of nucleophilicity depends no doubt on the dominance of one factor in both the reference system and in the systems to which it may be successfully applied, or else there must be a high degree of collinearity between the various factors. The factors influencing nucleophilicity are, however, more complicated than indicated above. In fact Bunnett has listed no less than seventeen factors which influence nucleophilic reactivity!![70] This does not mean that the ultimate multiple regression will involve seventeen explanatory variables. Many of the factors are rather specific to particular situations and there may be a high degree of collinearity between some of the factors. Nevertheless Bunnett's seventeen factors give a fair impression of the problems to be overcome in developing a comprehensive system of correlation analysis for nucleophilicity that is in any way comparable in its success with what now obtains for solvent effects (Section 5.4).

In the space available we can give only limited examples of the factors which Bunnett lists, over and above those already mentioned. The most notable additional factor is the *alpha effect*.[47] The principal diagnostic for this is the occurrence of large positive deviations in Brönsted-type plots. The nucleophiles concerned usually have an electronegative atom containing one or more unshared pairs of electrons *adjacent* to the nucleophilic atom and it is believed that the enhanced reactivity is due to these electrons - hence the use of "alpha" to describe the "effect". The alpha effect nucleophiles include OCl^-, HO_2^-, N_3^-, NH_2OH and N_2H_4. While the alpha effect is observed in reactions with some substrates, it is not observed in reactions with others. The explanation of the alpha effect is thus subtle. Pearson discussed the situation in 1972,[19] and Grekov and Veselov recently reviewed the topic comprehensively.[71] It remains to be seen whether it is possible to parametrize the alpha effect for correlation analysis.

Other factors mentioned by Bunnett include steric effects, ion-pairing, London and dipole attractions, and various hydrogen-bonding effects.

In these circumstances many workers seem to despair of the possibility of achieving any reasonably comprehensive system of correlation analysis for nucleophilicity. They either concentrate on the application of a comprehensive qualitative scheme[19] (usually HSAB, Section 6.3.4) or on the development of theoretical treatments for particular limited systems. Duboc surveyed some of these approaches in 1978;[20] the work of Klopman and of Hudson has been particularly important.[72] See also the work of Albery on the application of the Marcus relation to nucleophilic substitution[73] and the work of McLennan on the calculation of free energies of activation for Finkelstein reactions by a quasi-thermodynamic cycle.[74]

Some mention has already been made of gas-phase studies of nucleophilicity as a means of elucidating nucleophilic reactivity in the absence of solvation.[35] We give references to some recent studies.[75,76]

We now mention briefly certain other work on nucleophilic substitution which has not found a place earlier in this Chapter. Peterson *et al.* have developed a nucleophilicity scale based on the relative rates of reaction of pentamethyleneiodonium hexafluoro-antimonate with aqueous solutions of nucleophiles, equation (6.48);[77] *(see page 203)*. The scale encompasses both anions and uncharged nucleophiles and is related linearly to one based on the reactions of MeI in water. "It is inferred that the charge type of both nucleophiles and substrates is unimportant for S_N2 reactions in water."

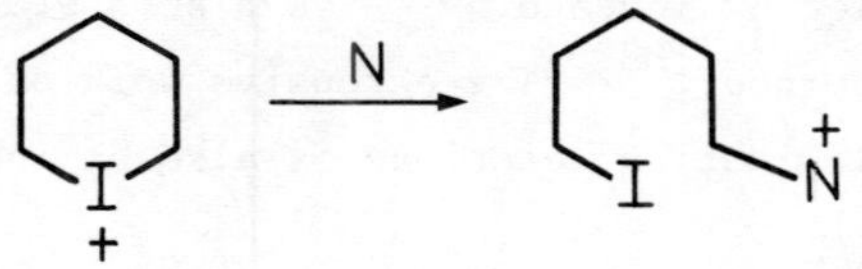

(6.48)

Quemeneur and Bariou have recently presented a comprehensive correlation analysis of about 400 rate constants for the reactions of amines, phosphines, and sulphides with alkyl halides in protic and aprotic solvents.[78] A multiparametric equation for variation of nucleophile, electrophile, and solvent has been developed in the form:

$$\log k_{(N+E)_S} = a\mathrm{B(S)} + b\mathrm{A(S)} + \alpha \qquad (6.49)$$

where a and A(S) relate to the electrophile, b and B(S) relate to the nucleophile and α is a factor, independent of solvent, which takes into account (among other influences) steric effects. The various parameters were tabulated and an impressive plot of log k (calc.) *versus* log k (obs.) included several hundred points. Arcoria *et al.* have also examined solvent effects on nucleophilic reactivity correlations and have introduced an S parameter (analogous to Ritchie's N_+) to deal with the role of the solvent in connection with multiple regressions.[79]

One aspect of the correlation analysis of nucleophilic substitution which has not so far been mentioned involves the study of leaving-group effects. This is, so to speak, the other side of the coin from the study of nucleophilicity. It involves studying the reactions of various electrophiles, differing only in leaving group, with a given nucleophile. One might expect some connection between nucleophilicity and leaving group ability: good nucleophiles should be poor leaving groups, and poor nucleophiles should be good leaving groups. Naturally such a broad generalization is hopelessly inadequate in detailed application. However, in favourable circumstances a linear Brönsted-type plot can be obtained, with a negative slope giving a β_{lg} value. We mention recent work of Jencks.[80,81] The correlation analysis of leaving group ability has also been much studied by Todesco.[82] See also a study by E. S. Lewis and Vanderpool.[83] The extensive work of Stirling and his colleagues on elimination reactions is also of interest in this connection.[84]

We have already mentioned that the correlation analysis of nucleophilicity features prominently in discussions of the reactivity—selectivity principle. We refer to some recent experimental contributions[85,86] and reviews[87,88] of this topic. Finally, we mention a review of "The Lewis Acid-Base Definitions: A Status Report" which is of interest from the standpoint of the correlation analysis of nucleophilicity.[89]

6.4 REFERENCES

1 R. P. Bell, *Acid—Base Catalysis*, Clarendon Press, Oxford, 1941.

2 R. P. Bell, *The Proton in Chemistry*, 2nd ed., Chapman and Hall, London, 1973.

3 E. F. Caldin and V. Gold, editors, *Proton-Transfer Reactions*, Chapman and Hall, London, 1975.

4 J. Hine, *Physical Organic Chemistry*, McGraw-Hill, New York, 1962.

5 L. P. Hammett, *Physical Organic Chemistry*, 2nd ed., McGraw-Hill, New York, 1970.

6 T. H. Lowry and K. S. Richardson, *Mechanism and Theory in Organic Chemistry*, Harper and Row, New York, 1976.

7 M. M. Kreevoy and R. W. Taft, *J. Amer. Chem. Soc.*, *77*, 5590 (1955), and references therein.

8 Ref. 1, p.135.

9 A. Buckley, N. B. Chapman, M.R.J. Dack, J. Shorter, and H. M. Wall, *J. Chem. Soc. (B)*, 631 (1968).

10 R. A. More O'Ferrall, W. K. Kwok, and S. I. Miller, *J. Amer. Chem. Soc.*, *86*, 5553 (1964).

11 J. N. Brönsted and K. Pedersen, *Z. Phys. Chem.*, *108*, 185 (1924).

12 J. N. Brönsted, *Om Syre- og Basekatalyse*, University of Copenhagen, 1926; English translation, *Chem. Rev.*, *5*, 231 (1928).

13 R. P. Bell and W.C.E. Higginson, *Proc. Roy. Soc. A*, *197*, 141 (1949).

14 R. P. Bell, in *Correlation Analysis in Chemistry: Recent Advances*,

N. B. Chapman and J. Shorter, editors, Plenum, New York, 1978, Chapter 2. (This is a review of the first half-century of the Brönsted equation.)

15 K. Bowden, N. B. Chapman, and J. Shorter, *Canad. J. Chem.*, *42*, 1979 (1964).

16 H. H. Jaffé, *Chem. Rev.*, *53*, 191 (1953).

17 E. F. Caldin, *Fast Reactions in Solution*, Blackwell, Oxford, 1964.

18 M. Eigen, *Angew. Chemie Int. Edn. in Engl.*, *3*, 1 (1964).

19 R. G. Pearson, in *Advances in Linear Free Energy Relationships*, N. B. Chapman and J. Shorter, editors, Plenum, London, 1972, Chapter 6.

20 C. Duboc, in *Correlation Analysis in Chemistry: Recent Advances*, N. B. Chapman and J. Shorter, editors, Plenum, New York, 1978, Chapter 7.

21 Examples in ref. 20, p.320.

22 F. J. Stubbs and C. N. Hinshelwood, *J. Chem. Soc.*, S71 (1949).

23 W. P. Jencks and J.Carriulo, *J. Amer. Chem. Soc.*, *82*, 1778 (1960).

24 R. G. Pearson, H. Sobel, and J. Songstad, *J. Amer. Chem. Soc.*, *90*, 319 (1968).

25 M. R. Crampton, *J. Chem. Soc.*, *(B)*, 2112 (1971).

26 M. R. Crampton and M. J. Willison, *J. Chem. Soc. Perkin Trans. II*, 238 (1974).

27 F. G. Bordwell and D. L. Hughes, *J. Org. Chem.*, *45*, 3314 (1980).

28 P. M. Bond, E. A. Castro, and R. B. Moodie, *J. Chem. Soc. Perkin Trans. II*, 68 (1976).

29 D. J. Hupe and W. P. Jencks, *J. Amer. Chem. Soc.*, *99*, 451 (1977).

30 D. J. Hupe, D. Wu, and P. Shepperd, *J. Amer. Chem. Soc.*, *99*, 7659 (1977).

31 C. G. Swain and C. B. Scott, *J. Amer. Chem. Soc.*, *75*, 141 (1953).

32 I. D. Biggs and D.L.H. Williams, *J. Chem. Soc. Perkin Trans. II*, 107 (1975).

33 G.-A. Craze, A. J. Kirby, and R. Osborne, *J. Chem. Soc. Perkin Trans. II*, 357 (1978).

34 Y. Ogata and Y. Yamauchi, *Bull. Chem. Soc. Japan*, *52*, 2171 (1979).

35 L. B. Young, E. Lee-Ruff, and D. K. Bohme, *J. Chem. Soc. Chem. Comm.*, 35 (1973).

36 C. L. Liotta, E. E. Grisdale, and H. P. Hopkins, *Tetrahedron Lett.*, 4205 (1975).

37 J. F. Evans and H. N. Blount, *J. Amer. Chem. Soc.*, *100*, 4191 (1978).

38 M. Cusumano, G. Faraone, V. Ricevuto, R. Romeo, and M. Trozzi, *J. Chem. Soc. Dalton Trans.*, 490 (1974).

39 Ref. 20, p.326.

40 J. O. Edwards, *J. Amer. Chem. Soc.*, *76*, 1540 (1954).

41 J. O. Edwards, *J. Amer. Chem. Soc.*, *78*, 1819 (1956).

42 T. Austad, *Acta Chem. Scand. A*, *29*, 895 (1975); *31*, 93 (1977).

43 L. H. Skibsted and J. Bjerrum, *Acta Chem. Scand. A*, *31*, 155 (1977).

44 B.-F. Liang and C.-S. Chung, *Inorg. Chem.*, *19*, 572 (1980).

45 M. Tabata and M. Tanaka, *J. Inorg. Nucl. Chem.*, *38*, 1529 (1976).

46 M. Tabata and M. Tanaka, *Inorg. Chem.*, *17*, 2779 (1978).

47 J. O. Edwards and R. G. Pearson, *J. Amer. Chem. Soc.*, *84*, 16 (1962).

48 K. M. Ibne-Rasa, *J. Chem. Educ.*, *44*, 89 (1967).

49 J. O. Edwards, *J. Chem. Educ.*, *45*, 386 (1968).

50 R. G. Pearson, *J. Chem. Educ.*, *45*, 581, 643 (1968).

51 C. D. Ritchie, *Accts. Chem. Res.*, *5*, 348 (1972).

52 C. D. Ritchie, *Pure and Appl. Chem.*, *50*, 1281 (1978).

53 Part XV of Cation—Anion Combination Reactions is C. D. Ritchie and M. Sawada, *J. Amer. Chem. Soc.*, *99*, 3754 (1977).

54 C. D. Ritchie, *J. Amer. Chem. Soc.*, *97*, 1170 (1975).

55 C. D. Ritchie and P.O.I. Virtanen, *J. Amer. Chem. Soc.*, *95*, 1882 (1973).

56 A. Pross, *J. Amer. Chem. Soc.*, *98*, 776 (1976).

57 Z. Rappoport, *Tetrahedron Lett.*, 2559 (1979).

58 P.O.I. Virtanen and R. Korhonen, *Acta Chem. Scand.*, *27*, 2650 (1973).

59 K. Hillier, J.M.W. Scott, D. J. Barnes, and F.J.P. Steele, *Canad. J. Chem.*, *54*, 3312 (1976).

60 C. A. Bunton, N. Carrasco, F. Davoudzadeh, and W. E. Watts, *J. Chem. Soc. Perkin Trans. II*, 1520 (1980).

61 J. L. Kice, T. E. Rogers, and A. C. Warheit, *J. Amer. Chem. Soc.*, *96*, 8020 (1974).

62 J. L. Kice and L. F. Mullan, *J. Amer. Chem. Soc.*, *98*, 4259 (1976).

63 E. G. Sander and W. P. Jencks, *J. Amer. Chem. Soc.*, *90*, 6154 (1968).

64 M. J. Cho and I. H. Pitman, *J. Amer. Chem. Soc.*, *96*, 1843 (1974).

65 I. H. Pitman, M. J. Cho, and G. S. Rork, *J. Amer. Chem. Soc.*, *96*, 1840 (1974).

66 J. P. Guthrie, *Canad. J. Chem.*, *52*, 2037 (1974).

67 G. Fischer, G. Küllertz, and A. Schellenberger, *Tetrahedron*, *32*, 1503 (1976).

68 J. P. Guthrie, *Canad. J. Chem.*, *56*, 962 (1978).

69 A. Finiels and P. Geneste, *J. Org. Chem.*, *44*, 1577 (1979).

70 J. F. Bunnett, *Ann. Rev. Phys. Chem.*, *14*, 271 (1963).

71 A. P. Grekov and V. Ya. Veselov, *Usp. Khim.* *47*, 1200 (1978); *Russ. Chem. Rev.*, *47*, 631 (1978).

72 See references collected in Ref. 20.

73 W. J. Albery, *Ann. Rev. Phys. Chem.*, *31*, 227 (1980).

74 D. J. McLennan, *Austral. J. Chem.*, *31*, 1897 (1978).

75 M. J. Pellerite and J. I. Brauman, *J. Amer. Chem. Soc.*, *102*, 5993 (1980).

76 S. P. McManus, *J. Org. Chem.*, *46*, 635 (1981).

77 P. E. Peterson, D. W. Vidrine, F. J. Waller, P. M. Henrichs, S. Magaka, and B. Stevens, *J. Amer. Chem. Soc.*, *99*, 7968 (1977).

78 F. Quemeneur and B. Bariou, *J. Chem. Res.*, (S) 187, 188, (M) 2344, 2357 (1979).

79 A. Arcoria, F. P. Ballistreri, and G. A. Tomaselli, *Tetrahedron*, *34*, 2545 (1978).

80 M. J. Gresser and W. P. Jencks, *J. Amer. Chem. Soc.*, *99*, 6963 (1977).

81 B. L. Knier and W. P. Jencks, *J. Amer. Chem. Soc.*, *102*, 6789 (1980).

82 G. Bartoli and P. E. Todesco, *Accts. Chem. Res.*, *10*, 125 (1977).

83 E. S. Lewis and S. H. Vanderpool, *J. Amer. Chem. Soc.*, *100*, 6421 (1978).

84 D. R. Marshall, P. J. Thomas, and C.J.M. Stirling, *J. Chem. Soc. Perkin Trans. II*, 1898 (1977).

85 E. Buncel and C. Chuaqui, *J. Org. Chem.*, *45*, 2825 (1980).

86 E. M. Arnett and R. Reich, *J. Amer. Chem. Soc.*, *100*, 2930 (1978).

87 C. D. Johnson, *Chem. Rev.*, *75*, 755 (1975).

88 C. D. Johnson, *Tetrahedron*, *36*, 3461 (1980).

89 W. B. Jensen, *Chem. Rev.*, *78*, 1 (1978).

CHAPTER 7
Some Further Statistical Matters

	Page
7.1. Regression: with or without intercept?	209
7.2. Criteria of goodness of fit	213
7.3. The relative importance of different 'effects' as indicated by multiple regression	220
7.4. References	225

In Chapter 2 the basic principles of simple and multiple linear regression were treated in a fairly informal way, as necessary background to the correlation analysis of organic reactivity. In the later Chapters occasional reference has been made (often in footnotes) to various issues connected with regression procedures, criteria of goodness of fit, and the interpretation of regression coefficients. The most important of these matters are concerned with the question of whether or not an intercept term should be an obligatory part of regression, various measures of goodness of fit which have been suggested for correlation analysis in organic chemistry, and the estimation of the relative importance of different 'effects' as indicated by multiple regression.[1]

7.1. REGRESSION: WITH OR WITHOUT INTERCEPT?

In Section 2.1 the procedure of simple linear regression was discussed in terms of the general equation for the straight line:

$$Y = aX + b \tag{7.1}$$

where a is the slope of the line and b is the intercept on the Y axis.

Later, in Section 2.2, multiple linear regression was discussed in terms of:

$$Y = a_1X_1 + a_2X_2 + a_3X_3 \ldots\ldots + b \qquad (7.2)$$

In these discussions it was assumed that the calculation of the intercept term b was an integral part of regression procedures.

In Chapter 3 the Hammett equation was introduced in the forms:

$$\log k = \log k^{o} + \rho\sigma \qquad (7.3)$$

$$\log K = \log K^{o} + \rho\sigma \qquad (7.4)$$

and it was said that "the symbol k^o or K^o denotes the statistical quantity (intercept term) approximating to k or K for the 'parent' or 'unsubstituted' compound". In a footnote it was then stated that "the form of the equation is subject to minor variations, of which the most serious is the combining of the log terms as log (k/k^o) or log (K/K^o), with k^o or K^o now referring to the *observed* value for the parent system". It was acknowledged that this procedure is necessary in the definition of σ from benzoic acid ionization (or, for that matter, in the definition of any other scale of substituent constants) but it was asserted that "it is not a necessary general procedure to force the regression through the point for the parent system. It is preferable to confirm that the intercept term is indistinguishable from the experimental value for the parent system and thus check that there is no effect connected with the replacement of H by *any* other substituent. If only *relative reactivities* are available, then log (rel. react.) is the dependent variable .... but the regression should still not be forced through the origin. An intercept should be permitted in the least squares treatment but in the ordinary way should emerge as indistinguishable from zero".

The issue came up again in the context of multiple regression. In Section 3.3, Taft's DSP equation was presented in the form:

$$P^i = \rho_I^i\sigma_I + \rho_R^i\sigma_R \qquad (7.5)$$

in which P^i is the substituent-influenced property, expressed relative to hydrogen as substituent, *e.g.* for chemical reactivity

$P^i = \log (k/k^o)_i$ or $\log (K/K^o)_i$; for spectroscopic frequencies $P^i = (\nu - \nu^o)_i$, *etc.* It was pointed out, in a footnote, that "Taft and his colleagues always express the correlated property relative to that for H, and no intercept term is allowed ...... Other authors, notably Charton, use the equation in the form:

$$Q = \alpha\sigma_{I,X} + \beta\sigma_{R,X} + h \qquad (7.6)$$

where Q is the property to be correlated, *not* expressed relative to hydrogen as substituent, and h is introduced as the appropriate intercept term. We may call this form the 'extended Hammett equation'. In the enormous variety of processes to which Charton has applied this equation, h is not always found to correspond closely to the value of the correlated property for H as substituent".

In Section 3.4 we saw how Swain and Lupton[2] analysed the various σ scales in terms of field and resonance parameters, $\mathcal{F}$ and $\mathcal{R}$ respectively, by means of the equation:

$$\sigma = f\mathcal{F} + r\mathcal{R} \qquad (7.7)$$

These authors, however, permitted an intercept term i in the multiple regressions "to avoid giving the unsubstituted compound infinite weight, to avoid forcing the line to go through the origin, since measurements on the substituted compound are in general no more accurate than those on substituted ones". (The phrase "infinite weight" seems to be a hyperbole for "excessive weight".) However, in most cases examined by Swain and Lupton, i turned out to be not significantly different from zero.

This, then, is the situation with regard to the intercept question as it has emerged so far in this book: there seems to be some disagreement as to the best procedure to follow. What more is there to say?

In the first place it must be said that statisticians tend not to approve of regression through the origin. Many textbook accounts of simple and multiple regression do not deal with regression through the origin at all, and those that do, contain stern warnings. For instance:[3] "Prior knowledge concerning the relationship sometimes requires that the regression equation should pass through the origin ..... then only the slope parameter has to be estimated ..... Great care must, however, be exercised in the adoption of this simplified model, since any bias in the measurements ..... would result in a biased regression coefficient and could lead to a false picture."

Since Ehrenson, Taft and their colleagues are the leading exponents of the use of regression through the origin in correlation analysis in organic chemistry it would be appropriate to summarize the arguments which they use to justify it.[4] They "recognize that the data for the unsubstituted (H) member of a set may be as subject to experimental error as any other member" but they hold such error to be "generally relatively small for a set of reliable data". Further, they claim that "any constant error from this source will be distributed among all of the substituents in such a manner as to achieve 'best fit' ". They believe that "any loss in precision of fitting of the set which may result from such a procedure is a small price to pay compared to the 'violence' done by the introduction ..... of a completely variable 'constant' parameter".

They continue: "The third parameter gives unwarranted freedom in the fitting of the data. In a data set, the point $P^i \equiv 0$ for the H substituent is a highly critical and valuable one in locating the scale origins. The hydrogen data point cannot influence the rho values (all sigmas are constrained to zero), and accordingly, the hydrogen data point is not explicitly counted in the statistics. Furthermore, the H atom has unique structure (perhaps the most different) among those of all the substituents. To in effect throw away this point in a data set (by introducing a third parameter) we believe is to render to the noise level one of nature's most revealing signals." They assert finally that many of the significant results of their analysis of reactivity in the benzene series (see Section 3.3) would have been lost had they not adhered strictly to equation (7.5).

For the correlation analysis of the behaviour of *meta*- and *para*-substituted benzene derivatives and closely analogous systems, we may be confident that little harm results from expressing measured properties relative to the unsubstituted system. Possibly, as Ehrenson and Taft appear to assert, there is some benefit, for purposes of comparison, in compelling the data to conform to equations without intercepts. It seems prudent, however, always to use the unconstrained equations initially, *e.g.* (7.3), (7.4), and (7.6), to check whether the intercept term approximates to the observed value for the unsubstituted system. Alternatively, following Swain and Lupton, the constrained equations such as (7.5) can be made unconstrained initially by allowing an intercept term. If this turns out to be not significantly different from zero, then the regression may safely be repeated with the constrained equation.

There are certainly many systems for which we cannot predict with confidence that the unsubstituted compound will behave as the effective parent of the substituted compounds. This is in fact just because "the H atom has a unique structure (perhaps the most different) among those of all the substituents", as we have quoted above.[4] In such cases the extended Hammett equation (7.6) (or similar equations for other purposes) must be applied and we shall often find that the intercept term does not approximate to the observed value for the H compound. This is the situation, in particular, for the correlation analysis of the *ortho*-effect and other cases in which a steric effect term must be added to (7.6), (see Chapter 4).

7.2. CRITERIA OF GOODNESS OF FIT

In Chapter 2 we presented the primary criteria of goodness of fit in regression: the *standard deviation of the estimate*, s, and the *correlation coefficient*, r for simple regression and R for multiple regression. We were at pains to emphasize that s measures the effectiveness of the regression equation as a summary of the data, while r or R is relevant to the success of the regression in providing an 'explanation' of the data, *i.e.* as testing the validity of a model. We also emphasized the need to assess correlation coefficient in relation to the degrees of freedom of the regression, *i.e.* the importance of an adequate confidence level. The problem of comparing the precisions of regressions involving different numbers of data points was also discussed in terms of Wold's C statistic (Section 2.1). The role of standard errors of the regression coefficients was also emphasized, particularly in relation to the problem of collinearity in multiple regression.

We mentioned, in passing, that certain physical organic chemists have devised other measures of precision in linear regression. We must now discuss these and various related matters.

First we refer to two articles by physical organic chemists which deal with the problem of criteria of goodness of fit. C.K. Hancock (1965) wrote on "Some Misconceptions of Regression Analysis in Physical Organic Chemistry".[5] He pointed out the limitations of r when one is comparing two regressions with different numbers of degrees of freedom and also the limitations of s if treated as a unique measure of precision. Hancock did not, however, distinguish clearly between the summarizing and explanatory roles of regression, but he did suggest that "comparison of significance should be made on the basis of the estimated standard errors of the regression coefficients and based on the appropriate tests of significance".

Davis and Pryor (1976) wrote on "Measures of Goodness of Fit in Linear Free Energy Relationships".[6] Much of the article is devoted to discussing "the inadequacy of the correlation coefficient", the inadequacy being that "r depends on the magnitude of the slope". By this the authors mean that, for a series of lines with a fixed number of data points and a given value of s, the values of r increase with the slope. To the present author this does not seem to be an "inadequacy" of r, but to reflect the role of r as a measure of the success of the regression in explaining the data. The authors do, however, point out that systems giving low Hammett ρ values present a special difficulty. Inevitably the values of r are poor and in the limit of $\rho = 0$, r will be zero. The bald statistics then require us to say that there is no significant relationship between log k and σ. Such a situation would, in isolation, leave nothing to discuss. But, chemically speaking, we regard such a result as highly significant, and proceed to devise explanations in terms of the absence of polar character in the reaction under study, or the cancellation of opposed polar influences, *etc*. For physical organic chemists, $\rho = 0$ is part of the continuum of positive to negative ρ values.

As a result of their discussion of the problems connected with s and with r, Davis and Pryor advocated basing a measure of goodness of fit on the standard deviation of the regression coefficient, in association with the t distribution (see the article for details[6]).

We pass now to various other quantities which physical organic chemists have suggested for expressing goodness of fit. The main purpose in developing these is to get away from the misleadingly small numerical scale within which correlation coefficients commonly lie in correlation analysis in organic chemistry. We begin with the magnitude ψ proposed by Exner, which we will introduce in the context of simple linear regression.[7]

ψ is defined by equation (7.8).

$$\psi = \left[\frac{n(1 - r^2)}{n - 2}\right]^{\frac{1}{2}} \tag{7.8}$$

Its meaning may be better appreciated if we recall from Chapter 2 that:

$$r = \left[1 - \frac{RSS}{TSS}\right]^{\frac{1}{2}} \tag{7.9}$$

where RSS = residual sum of squares and TSS = total sum of squares. Thus:

$$\psi = \left[\frac{n(RSS)}{(n - 2)(TSS)}\right]^{\frac{1}{2}} \tag{7.10}$$

and we see that ψ essentially compares directly the scatter about the regression line with the scatter about the mean value of the dependent variable. For the perfect regression $\psi = 0$, and for a less than perfect regression ψ is > 0, and increases numerically the poorer the fit. Further, the factor $n/(n - 2)$ gives some consideration to the effect of degrees of freedom, *i.e.* it corrects for the misleading impression given by high correlation coefficients in regressions involving few data points.

Thus ψ is a kind of disguised correlation coefficient, on a numerical scale from 0 upwards rather than from 1.0 downwards. This transfer of scale involves considerable numerical expansion, *e.g.* for $n = 10$, $r = 1.00$ to 0.90 becomes $\psi = 0.00$ to *ca.* 0.49, and correspondingly for $n = 4$, $\psi = 0.00$ to *ca.* 0.62. It has been suggested that values of ψ should be appraised as follows:[7]

<0.02	Very good
0.02 — 0.10	Good
0.10 — 0.20	Fair
0.20 — 0.50	Poor
0.50	Utmost admissible limit

It is claimed that the greater sensitivity of ψ makes it a clearer indicator than r as to what a correlation really accomplishes. Exner gives a revealing example based on a Hammett correlation for which $r = 0.956$ and $n = 12$.[8] By most standards this would be deemed a satisfactory or even good correlation. But $\psi = 0.32$, which roughly means that the measured values of log k are represented by the Hammett equation with a standard deviation of about 30 per cent of that achieved by the simple assumption that the substituents have no effect on the reactivity!!

It must be mentioned that in its incorporation of a term to correct for the effect of degrees of freedom, ψ bears a resemblance to the quantity C based on a "coefficient of determination" C^2 which was used by Swain and Lupton[2] instead of multiple correlation coefficient, R.† Their equation for C^2, written with our symbols and for simple regression, would be (7.11):

$$C^2 = \left[1 - \frac{(n-1)(RSS)}{(n-2)(TSS)}\right] \tag{7.11}$$

The quantities $RSS/(n-2)$ and $TSS/(n-1)$ are described as

† Swain and Lupton's C should not be confused with Wold's C statistic referred to above (see Section 2.1).

population-corrected squares of standard deviations of observed values of the dependent variable from values calculated from the regression equation and from the mean experimental value respectively. We may compare equation (7.11) with equation (7.9) for r. The correction $(n - 1)/(n - 2)$ is less pronounced than the correction $n/(n - 2)$ in equation (7.10) for ψ^2, and of course C^2 is still on essentially the same numerical scale as r^2. C^2 is always less than r^2, but as $n \rightarrow \infty$, $C^2 \rightarrow r^2$; also, as $n \rightarrow \infty$, $\psi^2 \rightarrow 1 - r^2$ and $\psi^2 \rightarrow 1 - C^2$. Incidentally, C^2 is commonly represented by statisticians as $\bar{r}^2$ (or $\bar{R}^2$ for multiple regression) and is referred to as r^2 or R^2 "adjusted for degrees of freedom".

For multiple regression on $m - 1$ explanatory variables, ψ is given by the following equations:

$$\psi = \left[\frac{n(1 - R^2)}{n - m}\right]^{\frac{1}{2}} \tag{7.12}$$

$$\psi = \left[\frac{n(RSS)}{(n - m)(TSS)}\right]^{\frac{1}{2}} \tag{7.13}$$

The corresponding equation for C^2 is:

$$C^2 = \left[1 - \frac{(n - 1)(RSS)}{(n - m)(TSS)}\right] \tag{7.14}$$

A statistic which is somewhat similar to ψ has been introduced by Ehrenson and Taft.[9,10] It is denoted by f and is defined by them as SD/RMS, where SD is the standard deviation of the estimate and RMS signifies the "root mean square of the data". It is important to realise that this definition of f is very much linked to these authors' practice of carrying out both simple and multiple regression without permitting an intercept term, as we have seen in Section 7.1. We must now clarify what is implied in the least squares procedure without an intercept, as compared to applying the method with an intercept.

While the usual procedure for a regression of Y or X is often regarded as 'unconstrained', this is not strictly true, for the regression must pass through the point $\bar{X}$, $\bar{Y}$. However, if no intercept is allowed, the procedure constrains the regression to pass through the origin, $X = Y = 0$. The equation for the slope is still:

$$a = \frac{\Sigma xy}{\Sigma x^2} \qquad (7.15)$$

but now x and y are identical to X and Y respectively, and are not given by $x = X - \bar{X}$ and $y = Y - \bar{Y}$. So the difference between the constrained and unconstrained regressions is that the former seeks to explain the variation in Y about $Y = 0$, and the latter to explain the variation in Y about $\bar{Y}$. The *TSS* and *ESS* of the constrained regressions are calculated with respect to $Y = 0$ and not from $\bar{Y}$; *RSS* is not affected. Other formulae connected with regression are correspondingly changed. Thus:

$$s = \left[\frac{RSS}{n - 1}\right]^{\frac{1}{2}} \qquad (7.16)$$

cf. the $n - 2$ in the denominator in the expression for s in unconstrained regression: the number of degrees of freedom is now $n - 1$ and not $n - 2$. The expression for r looks the same as (7.9), but *TSS* is with reference to $Y = 0$.

We can now appreciate the significance of the quantity $f = SD/RMS$. It might have been expected that for *SD* the quantity s as defined by equation (7.16) would be taken. However Ehrenson and Taft actually obtain *SD* by dividing *RSS* by n and not $(n - 1)$. It should be noted, however, that the data point for the unsubstituted compound is not counted in n. (This is evident from a study of the Tables in Ehrenson and Taft's papers.[9]) *RMS* is $[TSS/n]^{\frac{1}{2}}$, so that f is given by:

$$f = \frac{SD}{RMS} = \left[\frac{RSS/n}{TSS/n}\right]^{\frac{1}{2}} = \left[\frac{RSS}{TSS}\right]^{\frac{1}{2}} \qquad (7.17)$$

Comparison with equation (7.10) shows that f is effectively an equivalent for constrained regression of ψ for unconstrained regression, but f and ψ are not identical as is sometimes implied.[8,11]

Although we have introduced f in the context of simple linear regression, it has been employed in practice mainly in connection with the DSP equation. (Division by n is used in calculating *SD*, irrespective of the number of degrees of freedom.) Ehrenson and Taft appraise values of f very much along the lines suggested above for ψ. Thus "experience has shown that correlations of good precision are those for which $SD/RMS \leqslant 0.1$".[9] It should be pointed out, however, that the scales of f and of ψ are really only comparable for the treatment of data sets in which the values of the dependent variable are fairly evenly distributed above and below the value for the parent system. If this condition holds, the mean value of the dependent variable will be close to that for the parent system and *TSS* for the constrained regression will be comparable to *TSS* for the unconstrained procedure. This is no doubt the case for many of the data sets treated by Ehrenson *et al*. However if, for example, the values of the independent variable are all higher than the value for the parent system, *TSS* for the constrained regression will be much larger than for the unconstrained, and f and ψ will not be on comparable numerical scales.

It must be emphasized again that the statistic $f = SD/RMS$ applies only to the constrained regression of data expressed relative to the parent system.[12] It is meaningless to calculate the *RMS* of data which are not so expressed, *e.g.* to calculate the *RMS* of log k values rather than of log (k/k^o). A defence and elaboration of the use of f as a criterion of goodness of fit has recently been given by Ehrenson.[13]

Finally it should be mentioned that Koppel and Palm suggested that another criterion of goodness of fit would be the standard deviation of the estimate expressed as a percentage of the range covered by the dependent variable in the data set. They use the symbol $s\%$ and have used this criterion in solvent effect correlations.[14] In a way this procedure applies rather directly the idea that was used to introduce correlation coefficient in Chapter 2, *i.e.* that "a correlation for which $s = 0.1$ units is highly meritorious if the values of Y cover, say 100 units, but is of little value if Y itself covers only 0.2 units". The use of $s\%$ has not, however, been followed by many authors.

7.3. THE RELATIVE IMPORTANCE OF DIFFERENT 'EFFECTS' AS INDICATED BY MULTIPLE REGRESSION

When we encounter a multiparametric equation which is said to be successful in correlating a set of reactivity data, it is natural to inquire into the physical significance of the various terms in relation to the reaction mechanism. Numerous examples of this were given in earlier Chapters. It is often tempting to inquire further whether it is possible to rank the terms in an order of importance, or to identify one term that effectively dominates the others, or even to quantify the relative importance of different 'effects'. To show how such queries may be answered requires considerable discussion.

We begin by recalling a few instances from earlier parts of this book in which we commented in various ways on the relative importance of the different effects represented by the explanatory variables. Thus in connection with the DSP equation in Section 3.3.1. the ratio ρ_R^i/ρ_I^i was defined as a blending constant λ, and in selected examples the values of λ were rationalized in terms of substrate structure and the nature of the process under study. Similarly, when discussing the Swain—Lupton treatment in Section 3.4.1, we referred to the assessment of the 'percent resonance' character of the various series through an expression (presented below as equation 7.18), in which the regression coefficients r and f are weighted by scale factors ϕ and ψ "to correct for the different numerical ranges of the $\mathcal{F}$ and $\mathcal{R}$ scales".

$$\%\mathcal{R} = \frac{100\psi|r|}{\phi|f| + \psi|r|} \qquad (7.18)$$

In discussing several examples of the Koppel—Palm treatment of solvent effects in Section 5.4.1, we referred to the contribution of certain terms in the equation as being "important" or "dominant", without specifying very clearly the criteria underlying such judgments. We must now try to clarify the criteria and examine the status of the various ways in which the relative importance of different effects is commonly assessed.

It should be said at the outset that the relative values of the regression coefficients do not, strictly speaking, give an immediate indication of the relative importance of the various effects. The latter are quantified in the regression equation not simply by the regression coefficients, but by the *products of regression coefficient and explanatory variable*, and the different explanatory variables may well have entirely different numerical scales. It follows that the relative importance of the various effects is liable to be quite different from the impression we may gain by a simple comparison of the regression coefficients. Further, the relative importance of the various effects is characteristic of the individual members of the data series, and may vary quite widely through the series. It may only be expressed for the series as a whole in the form of an average, from which individual members may deviate markedly.

A good example to illustrate these matters is provided by equation (7.19), previously given in Section 4.2.3, for the influence of polar and steric effects on the ionization of 13 arylaliphatic carboxylic acids:[15]

$$\log K_a = -6.045 + \underset{(\pm 0.023)}{2.666\sigma^*} + \underset{(\pm 0.003)}{0.252E_s} \qquad (7.19)$$

It might be supposed from the regression coefficients of σ^* and E_s that the steric effect is very much less important than the polar effect. However, the σ^* values used in the regression are in the range 0 to +0.6, while the E_s values are in the range 0 to -4.7. Thus the numerical scales of the explanatory variables differ greatly. To show the consequences of this, we may consider triphenylacetic acid, for which $\sigma^* = +0.56$ and $E_s = -4.68$. The corresponding terms in the regression are +1.49 and -1.18. Hence the polar and steric contributions to the change in acidity as between acetic acid (the standard) and triphenylacetic acid differ but little numerically, although they are opposite in sign. Other members of the series could be selected to show variously the dominance of the steric effect or of the polar effect.

It is, however, possible to give an *overall* assessment of such a situation by a procedure based on weighting each regression coefficient by the standard deviation of the corresponding explanatory variable.[16] In the case of equation (7.19) the weighting factors would be $[\Sigma(\sigma^* - \overline{\sigma^*})^2/12]^{\frac{1}{2}}$ and $[\Sigma(E_s - \overline{E_s})^2/12]^{\frac{1}{2}}$, giving values of 0.171 and 1.539. The overall percentage contribution from the polar effect may then be calculated as

$$100(0.171 \times 2.666)/[0.171 \times 2.666 + 1.539 \times 0.252],$$

that is 54%. The contribution from the steric effect is therefore 46%. Thus, in the series as a whole, the polar effect is only slightly more important than the steric effect.

The procedure used by Swain and Lupton is somewhat along these lines. They calculated scale factors for the regression coefficients of $\mathfrak{F}$ and $\mathfrak{R}$ with respect to the entire range of substituents, 42 in number, involved in their work (Section 3.4.1). The scale factors were based on mean deviations rather than root mean square deviations and they were applied throughout Swain and Lupton's work, even though most of the regressions involved only a limited selection of the 42 substituents. The use of mean deviations rather than root mean square deviations was said to "minimize the effect of extreme substituents such as N_2^+". These features of Swain and Lupton's weighting procedure mean that much significance should not be uncritically attached to small differences in values for %$\mathfrak{R}$ or %$\mathfrak{F}$ as between one regression and another.

We must now consider the significance of the blending constant $\lambda = \rho_R^i/\rho_I^i$ for Taft's DSP equation (Section 3.3.1) and the analogous quantities for the extended Hammett equation as used by Charton (see, for example, Section 4.5.4). All these quantities are commonly used without considering the different numerical scales of the various substituent parameters, σ_I, σ_R, *etc.* This limitation means that discussion of λ, *etc.* should be largely qualitative in character, and not be concerned with *minutiae*, unless the substituents involved in the regressions are broadly comparable.

The assessment of the relative importance of particular effects is a special problem in the correlation analysis of solvent influence, because the numerical scales of solvent parameters are extremely varied.[16] This is well illustrated by the Koppel—Palm treatment, equation (7.20), for the influence of 44 aprotic solvents on the reaction between diazodiphenylmethane and benzoic acid (see Table 5.3, 5 and related discussion in Section 5.4.1).

$$\log k = -3.13 + \underset{(\pm 0.37)}{4.58}\,\frac{\varepsilon-1}{2\varepsilon+1} + \underset{(\pm 1.6)}{11.96}\,\frac{n_D^2-1}{2n_D^2+1} + \underset{(\pm 0.020)}{0.195}E - \underset{(\pm 0.0008)}{0.0181}B \qquad (7.20)$$

Naïve comparison of the regression coefficients would suggest that the polarizability term involving n_D^2 is the most important, while the solvent basicity term involving B is the least. In fact quite the reverse is true, since values of $(n_D^2 - 1)/(2n_D^2 + 1)$ are concentrated in the range 0.17 to 0.26, while values of B range from 31 to 192. If it is desired to attribute definite percentages to the contributions made by the various effects, averaged over the whole reaction series, a procedure of weighting the regression coefficients, similar to that presented for equation (7.19), is appropriate. However, other ways of assessing the relative importance of the different effects are often adopted.

Kamlet and Taft essentially build this assessment into their regression procedure by normalizing their solvent parameters α, β, and π^* to the numerical scale 0—1.00 (Section 5.4.4). Thus the zero in all three cases is effectively given by saturated hydrocarbons, while α is taken as 1.00 for methanol, β as 1.00 for hexamethylphosphoramide, and π^* as 1.00 for dimethyl sulphoxide. The regression coefficients which emerge from multiple regression on these parameters are considered by Kamlet and Taft to give an immediate indication of the relative importance of solvent Lewis acidity, basicity, and polarization/polarizability for the process under analysis. From equation (7.21) for the reaction of DDM with benzoic acid in 44 aprotic solvents, the order of decreasing importance is β, α, π^*.

$$\log k = -0.2038 + \underset{(\pm 0.2092)}{2.7139}\alpha - \underset{(\pm 0.1216)}{3.7049}\beta + \underset{(\pm 0.1096)}{1.2134}\pi^* \qquad (7.21)$$

On dividing each coefficient by the |sum| of all three coefficients, the contributions of each effect on a percentage basis emerge as 49% (β), 35% (α), and 16% (π^*).

Not too much stress should be laid on such numerical estimates of relative importance. We need to remember that we are concerned with an analytical model, which has no absolute status. All that the reaction 'experiences' is an environment of electron clouds and nuclei; it is physical organic chemists who try to distinguish different kinds of effect.† Thus the main result emerging from the above is the dominance of specific (β and α) over non-specific (π^*) solvent effects.

Partly because of a certain scepticism as to the value of assessing numerically the relative contributions of different types of solvent influence, the author and his colleagues have adopted a different approach in their applications of the Koppel—Palm treatment.[17,18] Section 2.2 stressed the importance of stepwise regression, with calculation of regression coefficients, appropriate standard deviations, and correlation coefficient at each stage, along with partial correlation coefficients for the successive introduction of the various explanatory variables. Proper consideration of the correlation matrix was also urged. While these matters are particularly important in connection with the problem of collinearity, they also serve to indicate the order of importance of the various explanatory variables. Thus for the reaction of DDM with benzoic acid in 44 aprotic solvents, the Koppel—Palm parameters enter the regression in the order B, $(\varepsilon - 1)/(2\varepsilon + 1)$, E, and $(n_{\mathrm{D}}^2 - 1)/(2n_{\mathrm{D}}^2 + 1)$, the successive correlation coefficients for the one-, two-, three-, and four-parameter regressions being 0.809, 0.902, 0.940, and 0.976. We may take the order in which the parameters enter as the order of

† An analogous comment could be made about the quantification of the relative importance of inductive and resonance effects in the DSP or extended Hammett equations. The analysis of total substituent effect through σ_I and σ_R scales is not the only possibility. (See comment about Godfrey's work in Section 3.5.)

decreasing importance of the various effects they represent.

Further, the relative errors of the regression coefficients, *i.e.* s_{a_1}/a_1, s_{a_2}/a_2, *etc.*, in the final multiple regression also give the order of importance of the various effects: the most precisely defined regression coefficient indicates the most important effect. (The relative errors are actually the reciprocals of the t values used in assessing the significance levels of the regression coefficients.) For the reaction of DDM with benzoic acid, inspection of Table 5.3, 5 from this point of view confirms the order of decreasing importance as B, $(\varepsilon - 1)/(2\varepsilon + 1)$, E, and $(n_D^2 - 1)/(2n_D^2 + 1)$.

It is also possible to infer the relative importance of the various terms in the final equation by considering the generation of that equation from each of the several possible equations which lack one of the explanatory variables, *i.e.* for $Y = f(X_1, X_2, X_3, X_4)$, from $Y = f'(X_1, X_2, X_3)$ by introducing X_4; from $Y = f''(X_2, X_3, X_4)$ by introducing X_1 ..... *etc.* The partial correlation coefficient for the introduction of the additional explanatory variable gives a measure of the importance of that variable. In the case of the Koppel—Palm equation (7.20) the partial correlation coefficients for the introduction of B, $(\varepsilon - 1)/2\varepsilon + 1)$, E, and $(n_D^2 - 1)/(2n_D^2 + 1)$ into the respective three-parameter equations are 0.96, 0.89, 0.84, and 0.77 respectively, thereby confirming the order of importance as inferred from the other criteria.

7.4 REFERENCES

1 In this Chapter obvious references which have been cited in earlier Chapters will not always be repeated.

2 C. G. Swain and E. C. Lupton, *J. Amer. Chem. Soc.*, *90*, 4328 (1968).

3 O. L. Davies and P. L. Goldsmith, editors, *Statistical Methods in Research and Production*, Oliver and Boyd, Edinburgh, 4th edn., 1972, p.201. Quotation by kind permission of the Longman group.

4 S. Ehrenson, R.T.C. Brownlee, and R. W. Taft, *Progr. Phys. Org. Chem.*, *10*, 1 (1973). Quotation from p.38 by kind permission of the publisher, John Wiley and Sons Inc., New York.

5 C. K. Hancock, *J. Chem. Educ.*, *42*, 608 (1965).

6 W. H. Davis and W. A. Pryor, *J. Chem. Educ.*, *53*, 285 (1976).

7 O. Exner, *Coll. Czech. Chem. Comm.*, *31*, 3222 (1966).

8 O. Exner, in *Advances in Linear Free Energy Relationships*, N. B. Chapman and J. Shorter, editors, Plenum, London, 1972, p.19.

9 Ref. 4, p.3. Values of *f* are prominent in the numerous Tables in this article.

10 P. R. Wells, S. Ehrenson, and R. W. Taft, *Progr. Phys. Org. Chem.*, *6*, 147 (1968); see in particular p.165.

11 J. Shorter, *Correlation Analysis in Organic Chemistry*, Clarendon Press, Oxford, 1973, Appendix, p.107.

12 See comment in A. Cornélis, S. Lambert, P. Laszlo, and P. Schaus, *J. Org. Chem.*, *46*, 2130 (1981).

13 S. Ehrenson, *J. Org. Chem.*, *44*, 1793 (1979).

14 I. A. Koppel and V. A. Palm, in *Advances in Linear Free Energy Relationships*, N. B. Chapman and J. Shorter, editors, Plenum, London, 1972, Chapter 5; especially Tables therein.

15 K. Bowden, N. B. Chapman, and J. Shorter, *J. Chem. Soc.*, 3370 (1964).

16 *cf*. T. M. Krygowski and W. R. Fawcett, *J. Amer. Chem. Soc.*, *97*, 2143 (1975).

17 A. G. Burden, N. B. Chapman, H. F. Duggua, and J. Shorter, *J. Chem. Soc. Perkin Trans. II*, 296 (1978).

18 M. H. Aslam, G. Collier, and J. Shorter, *J. Chem. Soc. Perkin Trans. II*, 1572 (1981).

APPENDIX
Some Terminology of Physical Organic Chemistry

It is not intended to provide a glossary of all the technical terms of physical organic chemistry that are used in this book. The meanings of any terms unfamiliar to the reader should usually be sought in the various textbooks of physical organic chemistry which have been cited in the references at the end of each Chapter, or in the IUPAC glossary.[1] However, there is a variety of terminology which may be used in connection with electronic explanations of the effect of structure upon reactivity and there is considerable confusion. We shall now explain the approach to this terminology which was adopted for this book.[2]

The *polar effect* of a substituent comprises all the mechanisms by which the substituent may modify the electrostatic forces operating at a reaction centre, relative to a standard substituent, which is usually a hydrogen atom. These forces may be governed by charge separations arising from differences in the electronegativity of atoms (generating dipoles), the presence of unipoles, or electron delocalization. Field (or direct), inductive (through-bond polarization involving σ- or π-electrons), and conjugative effects may in principle be distinguished. Because of the difficulty of distinguishing between field and through-bond effects in practice, the term *inductive effect* is often used to cover both, and is normally so used in this book. (An alternative procedure is to use *localized effect* as the collective term.) For conjugative substituent effects the general term *resonance effect* is normally used in this book. This term has long been used in connection with correlation analysis and avoids making a distinction between mesomeric (time-permanent) and electromeric (time-variable) effects, which together make up the tautomeric or total conjugative effect. Objection is sometimes raised to the term 'resonance' on the grounds that it is liable to be misconstrued in terms of oscillation between discrete structures but if this error is avoided, the word is the most useful single term available to us. Thus the primary use of the term resonance effect is to cover conjugation of the substituent with the delocalized system, which may (see I) or may not (see II), include the functional group.

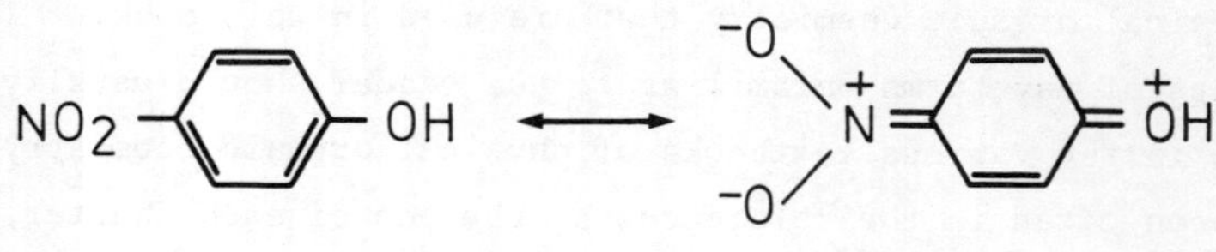

(I)

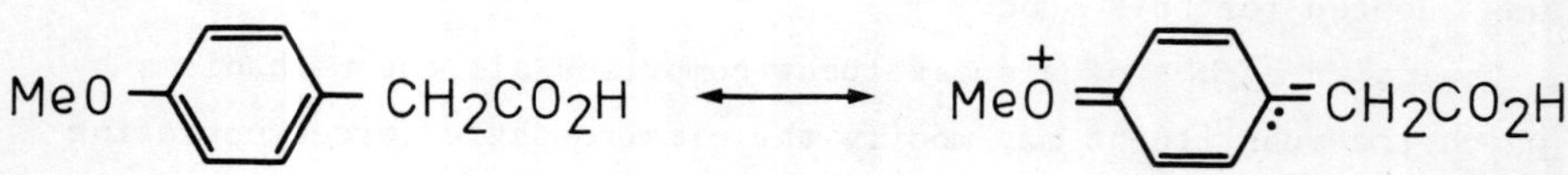

(II)

The term resonance effect is also used in connection with the conjugation of the functional group with the delocalized system, *e.g.* (III). It is used particularly in this sense in Chapter 4. The

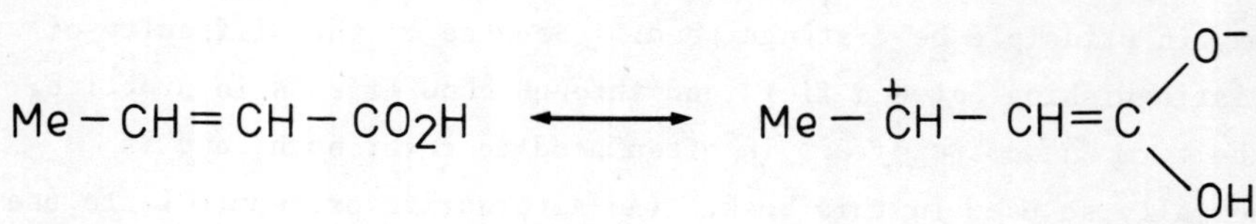

(III)

precise meaning of resonance effect may thus only be understood in context.

Hyperconjugation or the *hyperconjugative effect* is historically and commonly regarded as a special kind of resonance effect involving primarily delocalization of electrons in C—H bonds adjacent to an unsaturated system, *e.g.* (IV).

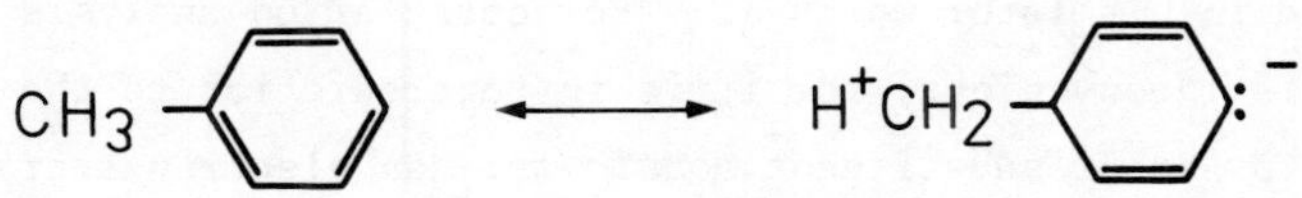

(IV)

The actual nature of this substituent effect, however, must be regarded as still in doubt. The term simply means a special effect of α-hydrogen atoms in the same sense as conjugative electron release. An analogous effect of C—C bonds is also sometimes invoked; this is C—C hyperconjugation, in contrast to C—H hyperconjugation. (See Section 4.3.1.)

Steric effects are caused by the intense repulsive forces which operate when two non-bonded atoms approach each other so closely that non-bonded compressions occur. A *primary steric effect* of a substituent is the direct result of the difference in compression energy which arises because the substituent differs in structure from a suitable standard substituent in the vicinity of the reaction centre. A *secondary steric effect* involves the moderation of a polar effect or resonance effect by non-bonded compressions.

We turn now to the question of symbols and signs for the electronic effects of substituents. In accordance with the choice of terms advocated above we have used I for the inductive effect and R for the resonance effect. In this book we have avoided describing electron delocalization by symbols such as M, E, T, C, and K, which are used to a greater or a less extent in many books and articles on physical organic chemistry.

The question of sign-convention for electronic effects is more difficult. The situation is tangled, as situations regarding sign convention often are. The Ingold sign convention is widely used in organic chemistry, all over the world.[3] Its basis lies in associating *electronegativity* (relative to the hydrogen atom) with a *negative* sign, *electropositivity* with a *positive* sign. Thus the nitro group is described as electron-withdrawing by virtue of its $-I$ and $-K$ effects (to use the symbol for the conjugative effect favoured by Ingold in his later works[4]). For correlation analysis this convention is inconvenient, for it is in contradiction to the sign convention for polar substituent *constants*: an electron-attracting group is given a *positive* substituent constant and an electron-releasing group is given a *negative* substituent constant (Section 3.1). Authors in the field of correlation analysis often avoid this contradiction by adopting the other sign convention for electronic effects, originally associated with Robinson.[5] This practice has been followed in this book; thus NO_2 is a $+I$, $+R$ substituent; Cl, a $+I$, $-R$ substituent, *etc.*[6]

REFERENCES

1 V. Gold, *Pure Appl. Chem.*, *51*, 1725 (1979).

2 This account is adapted from J. Shorter, *Correlation Analysis in Organic Chemistry*, Clarendon Press, Oxford, 1973, pp.5-7, by kind permission of the publisher.

3 C. K. Ingold, *Chem. Rev.*, *15*, 225 (1934).

4 C. K. Ingold, *Structure and Mechanism in Organic Chemistry*, Bell, London, 2nd edn., 1969.

5 R. Robinson, *An Outline of an Electrochemical (Electronic) Theory of the Course of Organic Reactions*, Institute of Chemistry, London, 1932, p.34.

6 The present author has been involved in a variety of practice. Ref. 2 uses the Robinson sign convention but see, however, the two monographs edited by N. B. Chapman and J. Shorter, to which frequent reference has been made in this book.

INDEX

This index is a guide to the main entries for important topics. Individual compounds, reactions, *etc.* are not usually mentioned. T after a page number indicates a Table.

Acid-base catalysis, 174-182

Acids, hard and soft, 193-195

Additivity of substituent effects, 62,83,87-90

Alkyl groups, significance of σ^*, 94-97,105-110

Alpha effect, 201

Anchimeric assistance, 39,85

Bases, hard and soft, 193-195

Brönsted equation, 1,176-182

Brönsted-type relationships, 182-186

Brown's σ^+, 31,32T,35-37

Brownstein's LFER, 144

Bunnett, factors influencing nucleophilicity, 201

Charton,
- application of extended Hammett equation, 48,115-118, 211,222
- correlation analysis for polycyclic arenes and heterocyclic systems, 63

Charton *(cont.)*
- criticism of Taft analysis, 105-110
- treatment of *ortho*-effect, 115-118
- treatment of steric effects, 103-107,105T

Chemical shifts, ^{13}C, 64,96,120; ^{19}F, 50; ^{1}H, 90,96

Coefficient of determination, C^2, 55,216

Collinearity, 22-24,155-158, 224

Confidence level, 15-18,20-22

Correlation analysis in organic chemistry, introduction, 1-6

Correlation coefficient, Chapter 2
- multiple, 20-23,216,224
- partial, 21,225
- simple, 12-18,30,213-217

Cross-conjugation, 31-35,228

Degrees of freedom in correlation, Chapter 2, 35,

Degrees of freedom in correlation *(cont.)*
214-217

DeTar, treatment of steric effects, 102,105-107

Dewar—Grisdale treatment, 63

Dielectric constant,
limitations, 136-138
treatment of solvent effects, 132-138,140T,147, 150T,155-159,166-168

Dimroth and Reichardt's E_T, 140T,143,148,154-161,168

Duality of substituent constants, 31-33

Dual substituent-parameter equation, 45-54,210-213,222

Dubois,
bromination of olefins, 88
treatment of steric effects, 100, 101T,105

Edwards equation, 191-193

Ehrenson's DSP equation, see dual substituent-parameter equation

Electronic effects:
terminology, symbols, and signs, 227

Electrophilic aromatic substitution, 42-44

Electrophilic solvation, 147-153

Exner,
analysis of σ values, 52-54, 53T
views on inductive effect,64
ψ statistic, 215-219

F distribution, 20

Field effect, 55,63-65,227

Fujita,
examination of E_s^c, 102
treatment of *ortho*-effect,118

Gas-phase reactivity, 65,95, 108,189,202

Godfrey's theoretical model, 67,118,224

Goodness of fit, measures of,
correlation coefficient, Chapter 2, 213-217
standard deviation of the estimate, 12-14,20,213-219
C (Wold), 17
C^2 (Swain and Lupton),55, 216
f (Ehrenson and Taft), 51, 217-219
ψ (Exner), 215-219
$s\%$ (Koppel and Palm), 219

Grob, σ_I^q substituent constant, 110, 111T

Grunwald and Winstein's Y, 139-142, 140T

Gutmann, DN and AN, 159,163

Hammett equation, 2, Chapter 3, 210; extended, 48,115-118, 211,222

Hancock, modified steric parameter, 91-94,92T,99, 101-105

Hard and soft acids and bases, HSAB, 193-195

Hughes and Ingold, treatment of solvent effects, 130-132

Hyperconjugation, 83-85,91-94, 99,101,105-107,229

Inductive effect, nature, 64, 227

Inductive and resonance effects, separation, 45-54

Jaffé
assessment of correlation coefficient, 15
multiparameter extension of Hammett equation, 62

Jaffé *(cont.)*
review of Hammett equation, 30

Jencks's equation, 199

Kamlet—Taft, treatment of solvent effects, 160-165, 163T,223
parameters, 162T

Kirkwood, dielectric function, 132-135,146-160

Koppel—Palm, treatment of solvent effects, 146-160, 151T,168,223-225
parameters, 150T

Kosower's Z, 140T,142

Krygowski and Fawcett, enthalpy and entropy analysis, 67,118
treatment of solvent effects, 159

Laidler and Eyring, treatment of solvent effects, 129, 132-136

Least squares, method of, Chapter 2, 28,217

Leaving groups, 203

Linear free-energy relationships (LFER), introduction, 1-6

Linear regression, Chapters 2 and 7,
multiple, 19-24,220-225
simple, 9-19
with or without intercept, 48,209-213,217-219

Menschutkin reaction, 128, 133-135,142,149,163

Mixed solvents, medium effects, 132-141

Molecular mechanics, 102

Multicollinearity, 22-24,155-158,224

Nucleophiles, hard and soft, 193-195

Nucleophilicity parameters, Chapter 6,
n (Swain—Scott), 186-191, 188T
n_{Pt}, 190,190T
n_{MeI}, 189,190T
H, E_{n}, P, (Edwards), 191-193, 192T
N_+, (Ritchie), 195-199,197T

Nucleophilic solvation, 147-153

Ortho-effect, 78,81,112-120

Ortho-substituent constants, 79T,114,114T

Palm, modified steric parameter, 93,92T,99

Pavelich—Taft equation, 86-90

Polarity parameters or scales; see Solvent polarity parameters

Reaction constant, Chapters 3 and 4,
ρ, 28,33T;167-169 (solvent effects)
ρ^*, 80-90
ρ_I and ρ_R, 47-54
$r^{\pm}$, see Yukawa—Tsuno equation

Reagent, effect on reactivity, Chapter 6

Refractive index, use as solvent parameter, 137, 147-158,150T

Regression coefficient, Chapter 2,
significance in relation to 'effects', 51,56,64,116, 220-225
standard error, 18,22,225

Regression, linear, see Linear regression

Resonance effect, Chapters 3 and 4, 227

Ritchie, work on nucleophilicity, 195-199

Significance level, 15-18, 20-22

Six-number, 92,101

Sliding scale of σ values, 32-35

Solvatochromic comparison method, see Kamlet— Taft

Solvent effects, Chapter 5,
  correlation analysis by substituent constants, 165-167
  early history, 128-130
  Hughes— Ingold treatment, 130-132
  Kamlet— Taft treatment, 160-165,223
  Koppel— Palm treatment, 146-160,168,223-225
  treatment through dielectric constant, 132-138, and see Dielectric constant
  treatment through solvent parameters, 138-144
  on Hammett ρ and σ, 167-170

Solvent parameters, Chapter 5,
  S (Brownstein), 144
  E_T (Dimroth— Reichardt), 140T,143,148,154-161,168
  Y (Grunwald— Winstein), 139-142,140T
  DN, AN (Gutmann), 159,163
  α, β, π*, (Kamlet— Taft), 160-165,162T
  Y, *P*, *B*, *E*, (Koppel— Palm), 146-160,150T
  Z (Kosower), 140T,142

Solvent polarity, concept, 130, 145

Standard deviation of the estimate, 12-14,20,213-219

Statistical methods, Chapters 2 and 7

Steric effects, Chapter 4, 229

Steric parameter,
  E_s, Chapter 4, 78T,79T,92T, 101T,221
  E_s^c, 91-94,92T,99,101-105
  E_s^o, 93,92T,99
  E_s', 100,101T
  ν, 104,105T

Student's *t* function, 16-18, 22,225

Substituent constants (polar), Chapters 3 and 4,
  σ, 30T,32T,32-35 (sliding-scale),34T (primary),169 (solvent effects)
  σ^-, 31,32T,36
  σ^+, 31,32T,35-37,42-45
  σ^n, 33,34T
  σ^o, 36,46,46T
  σ', 55,79
  σ_I, 45-54,46T,52-54 (scaling),53T (Exner),79,96, 108,108T,111T (Charton)
  σ_I^q, 110,111T
  σ_R, (or σ_R^o, σ_R(BA), σ_R^+, σ_R^-), 45-54,46T,52-54 (scaling), 53T (Exner),79
  σ_o, 78,79T,112,114T
  σ*, Chapter 4, 78T,168,221

Swain— Lupton treatment, 54-61, 118,211,220,222
  applications, 58
  critique, 59
  extensions, 60
  $\mathfrak{F}$ and $\mathfrak{R}$ values, 57T
  modification by Hansch, 58

Swain— Scott equation, 186-191

Symbols, for electronic effects 229

Taft's DSP equation, 45-54,210-213,222

Taft's polar and steric parameters, Chapter 4

Taft, separation of polar, steric, and resonance effects, 5, Chapter 4
  applications, 80-90
  development and critique to 1971, 90-99
  introduction, 74-80
  since 1971, 99-112

Unipolar substituents, 56,60, 65

van Bekkum *et al.*, sliding-scale of σ values, 32-35

Variable resonance parameter equation, 66

Wold's *C* statistic, 17

Yukawa and Tsuno,
  inductive (σ_i) and resonance (σ_π^+ and σ_π^-) parameters, 66

Yukawa—Tsuno equation, 35-45,
  applications, 37-40,38T
  critique, 40-45
  discussion of $r^\pm$, 37-42
  electrophilic aromatic substitution, 42-44
  nucleophilic reactions, 36, 38T,40
  relationship between $r^\pm$ and σ, 39
  significance of $r^\pm$ for two-step processes, 40